AF798933

Lecture Notes in Morphogenesis

Series Editor

Alessandro Sarti, CAMS Center for Mathematics, CNRS-EHESS, Paris, France

Advisory Editors

Henri Berestycki, Ecole des hautes etudes en sciences soci, Paris, France

Paul Bourgine, Ecole Polytechnique, Palaiseau, France

Giovanna Citti, Dipartimento di Matematica, Università di Bologna, Bologna, Bologna, Italy

Paolo Fabbri, Universitá di Venezia, Venzia, Italy

Vincenzo Fano, Dept. Basic Sciences & Foundations, Urbino University, Urbino, Italy

Sara Franceschelli, ENS, Lyon, France

Maurizio Gribaudi, EHESS, Paris, France

Annick LESNE, LPTMC UMR 7600, CNRS UMR 7600 case courrier 121, Paris cedex 05, France

Giuseppe Longo, Centre Cavaillès, Ecole Normale Supérieure, PARIS, France

Thomas Lorenz, Efiport GmbH, Frankfurt School Verlag, Frankfurt am Main, Hessen, Germany

Jean-Pierre Nadal, Service Facturier, EHESS, Paris, France

Nadine Peyriéras, Institut de Neurobiologie Alfred Fessard, CNRS-N&D, Gif sur Yvette, France

Jean Petitot, EHESS, Paris, France

Jan-Philip Schmidt, University of Heidelberg, Heidelberg, Germany

Carlos Sonnenschein, Dept Integrated Physio & Pathobiology, Tufts University School of Medicine, Boston, MA, USA

Ana Soto, Dept of IPP, Tufts University, Boston, MA, USA

Federico Vercellone, University of Turin, Turin, Italy

Lecture Notes in Morphogenesis is an interdisciplinary book series which aims to face the questions of emergence, individuation and becoming of forms from several different points of view: those of pure and applied mathematics, of computational algorithms, of biology, of neurophysiology, of cognitive and social structures. The set of questions above concerns all the manifestations of Being, all the manifestations of Life. At the heart of contemporary embryogenesis lies an essential question: How can form emerge from the constant, chaotic flow? How can a sequence of purely informational elements — an a-signifying combination of chemical substances organized in the DNA molecule — evolve into the highly complex and structured forms of the living organism? A similar question can be asked when we deal with the morphogenesis of vision in neural systems and with the creation of evolving synthetic images, since digital technology makes possible the simulation of emergent processes both of living bodies and of visual forms. Finally the very idea that abstract structures of meaning could be captured in terms of morphodynamic evolution opens the door to new models of semiolinguistics, semiotic morphodynamics, and cognitive grammars. An entire heritage of ideas and concepts has to be reconsidered in order to face new and challenging problems: the theoretical framework opened by Goethe with the introduction of the word "Morphogenesis" is developed by D'Arcy Thompson in "On Growth and Form", it is reorganized with new theoretical insights by the classical structuralism of Lévi-Strauss and formalized by the dynamical structuralism of René Thom. The introduction of the post-structuralists ideas of individuation (in Gilbert Simondon and Gilles Deleuze) and plasticity of structures builds a bridge to contemporary problems of morphogenesis at a physical, biological, social and transindividual level. The objective of this book series is to provide suitable theoretical and practical tools for describing evolutionary phenomena at the level of Free boundary problems in Mathematics, Embryogenesis, Image Evolution in Visual Perception, Visual Models of Morphogenesis, Neuromathematics, Autonomy and Self-Organization, Morphogenetic Emergence and Individuation, Theoretical Biology, Cognitive Morphodynamics, Cities Evolution, Semiotics, Subjectivation processes, Social movements as well as new frontiers of Aesthetics. To submit a proposal or request further information, please use the PDF Proposal Form or contact directly: *Dr. Thomas Ditzinger (thomas.ditzinger@springer.com)*

More information about this series at https://link.springer.com/bookseries/11247

Alessandro Sarti · Giovanna Citti ·
David Piotrowski

Differential Heterogenesis

Mutant Forms, Sensitive Bodies

Springer

Alessandro Sarti
CAMS Center for Mathematics
CNRS-EHESS
Paris, France

Giovanna Citti
Department of Mathematics
University of Bologna
Bologna, Italy

David Piotrowski
LIAS Linguistics, Anthropology, Sociolinguistics
CNRS-EHESS
Paris, France

ISSN 2195-1934 ISSN 2195-1942 (electronic)
Lecture Notes in Morphogenesis
ISBN 978-3-030-97796-2 ISBN 978-3-030-97797-9 (eBook)
https://doi.org/10.1007/978-3-030-97797-9

© Springer Nature Switzerland AG 2022
This work is subject to copyright. All rights are reserved by the Publisher, whether the whole or part of the material is concerned, specifically the rights of reprinting, reuse of illustrations, recitation, broadcasting, reproduction on microfilms or in any other physical way, and transmission or information storage and retrieval, electronic adaptation, computer software, or by similar or dissimilar methodology now known or hereafter developed.
The use of general descriptive names, registered names, trademarks, service marks, etc. in this publication does not imply, even in the absence of a specific statement, that such names are exempt from the relevant protective laws and regulations and therefore free for general use.
The publisher, the authors and the editors are safe to assume that the advice and information in this book are believed to be true and accurate at the date of publication. Neither the publisher nor the authors or the editors give a warranty, expressed or implied, with respect to the material contained herein or for any errors or omissions that may have been made. The publisher remains neutral with regard to jurisdictional claims in published maps and institutional affiliations.

This Springer imprint is published by the registered company Springer Nature Switzerland AG
The registered company address is: Gewerbestrasse 11, 6330 Cham, Switzerland

What interests me is not the rules, but the fact that the rules change. It's mostly a matter of change.

—Nanni Balestrini

Contents

Chapter 1
Introduction

1.1 The Problematic Dimension of Becoming

The question we are interested in deals with forms, the becoming of forms. We are interested in morphogenesis as in the spirit of twentieth-century French philosophy and particularly of Gilbert Simondon and Gilles Deleuze work. For Simondon becoming is the passage from a pre-individual intensive plane to the individuation of forms extended in space and time [7]. The individual at the centre of becoming, rather than a stable identity, is understood to be engaged in a continual process of individuation. And in fact, for Simondon, individuals are always engaged in such a process of individuation so that an identity can never exist as such. The becoming of the individual unfolds, furthermore, in relation to a field of forces composed of kinetic and dynamic as well as perceptive and affective forces. When these forces find a synchronicity, a common resultant, they induce a flow that will constitute a new form which is always unpredictable, indescribable, unheard-of theretofore. The individual, then, is always temporary: it is in its nature to continue to become other than itself in the process without beginning and without end that is individuation. This was Gilbert Simondon's manner of conceiving of genesis: a constellation of intensive elements that becomes a flux and then a form in a perennial individuation of life.

Gilles Deleuze, in turn, took up Simondon's idea of individuation and elaborated upon it. For Deleuze of Difference and Repetition [1] the becoming of forms is a passage from a virtual plane to an actual one. He reconsiders the Simondonian concept of individuation and equips this passage with a differential calculus, wherein the evolution of forms is understood as the solution of a differential problem. The idea of Leibniz's differential calculus is revitalized so that the becoming of forms is now viewed as the solution of a distribution of differential constraints that populate the virtual. The virtual, then, is a multiplicity of differential generators that are the genetic elements of every morphodynamics. Whereas Simondon retained a primary focus on individuals, Deleuze was more concerned with imagination. He was interested

© Springer Nature Switzerland AG 2022

A. Sarti et al., *Differential Heterogenesis*, Lecture Notes in Morphogenesis,
https://doi.org/10.1007/978-3-030-97797-9_1

in understanding how it is that thought emerges and in trying to understand what happens when a new idea appears. In either case—whether a question of individuals or of imagination—it is a matter of composing fields of force and hoping that they work, that is, that they are capable of integrating in such a manner as to unleash a new form or a new idea. In this context, "integration" retains the precise meaning given to it by the mathematicians of the differential calculus, starting from Leibniz -that is, finding the solution of a problem. Bodies are thus understood to be located within a problematic field, to which their becoming itself unfolds as a solution. The action of the differential furthermore is not abstract. Far from it. Rather, it is in the very nature of the flesh to accomplish such integration—that is, to actualize sensitive, perceptive, and affective fields of force. Without this instrument of the differential we would not even know how to understand our own experience of feeling, seeing, and thinking. Indeed, we wouldn't be able to account for our experience, the whole of which is made up of integrated fields. Only thus can we account for the fact that the colours we see are not Newton's but are Goethe's, that is, they are all intertwined with one another to such an extent that when a new colour enters the scene all the others change hue.

It is a field effect due to the integration process that makes each color dependent on every other and makes our experience reliant upon the differential character of perceptual phenomenality.

The becoming of forms, then, is the actualization of a distribution of differential constraints, which latter constitute the virtual. The virtual is always a multiplicity of differentials which are the intensive genetic elements of every morphodynamics. In this framework, to imagine a new form means to compose differential fields that may or may not give rise to integration. Composing such differential fields means looking for adjunct fields [1] that allow for integration. We cannot know in advance what will be result from such a process of integration—neither the dimension, nor the parameters of the space of possibility of a new form, since that space only emerges in combination with the newly emergent form itself.

This composition of heterogenous differential fields with a conjunctive logic of "and ... and ... and" gives rise to assemblages, which constitute the virtual elements for heterogeneous dynamics ([2]). More generally, composing includes a plurality of actions, including adding, subtracting, cutting, hacking, and modifying differential field assemblages.

Differently from both mathematical physics and structural morphodynamics, where the becoming of forms emerges from generators that are homogenous in space and time, the heterogeneous dynamics in question—called *heterogenesis*—introduces the possibility of the mutation of laws in space and time so as to overcome any homogeneity. The virtual can be recomposed and can thus instantiate a dynamics of the event whereby new spaces of possibility and new forms can be generated.

In contrast to the structural dynamics formalized by René Thom in Theory of Catastrophes [8], we currently lack a mathematical theory for such heterogeneous dynamics. The problem we pose ourselves in what follows is thus to understand if this idea of heterogenesis is only a beautiful metaphysics, or if, to the contrary, it could really be implemented in terms of a heterogeneous differential dynamics.

We will deal with precisely this question: How might we re-think the differential so as to produce the heterogenetic becoming of forms? In this book we would like to reconsider and deepen this question, which we began to consider in the article "Differential Heterogenesis and the Emergence of Semiotic Function" [6].

1.2 From Structuralism to Post-structural Dynamics

To situate this problem historically, we can begin by noting that the operation we are performing shares certain aspects with the one of René Thom and Jean Petitot in the 70s–80s, though it pertains to different materials and dynamics. In those years, the structural paradigm supported by Lévi-Strauss in cultural anthropology, Jakobson in phonology, Saussure in linguistics, Greimas in semiotics, and Tesnière in syntax, to name only a few, had already been brought to fulfilment, at least in its philosophical deployment. With respect to this elaboration, the intervention of Thom-Petitot has to be considered as a true translation of structuralism in epistemic terms. Dynamic structuralism in facts interprets the theory of structures by means of the theory of catastrophes of René Thom, in such a way that structures become dynamical devices suitably controlled. In this way the semiotic square of Greimas becomes a catastrophe with four possible stable dynamics, while the canonical formula of myth of Lévi-Strauss becomes a catastrophe with eight stable dynamics. In the framework of a morphodynamical modelization of the Saussurean sign, David Piotrowski proposed to account for the ≪intensive≫ and ≪extensive≫ semantic relations with a catastrophe with two stable dynamics. By means of the control of dynamics it is possible to move from one side to the other of the semiotic square, or to pass from one state to another in the dynamics of the myth, or, again, by including the signifiers in the control chain, from one signified to another. Effectively at stake is a sort of puppet opera wherein suitable parameters control the dynamics, just as wires suspended from the ceiling control the puppets in the theatre.

Up to here we are dealing with a dynamics with a strong possibility of control. But after Michel Foucault's harsh criticism of structuralism as a relational system of empty and interchangeable places, and thanks to the exceptional work of Deleuze and Guattari [2] over the course of the 80s, the idea of the becoming of forms was drastically transformed. And here enters the concept of heterogenesis as a device capable of instantiating infinite dynamics thanks to a virtuality in continuous recombination.

The oppositional relations of the structure are replaced by an assemblage of heterogeneous relations in continuous composition. Instead to choose between the oppositional terms of a structure it is now possible to compose assemblages of heterogeneous relations, which allows in turn for the emergence of transient dynamics and vibrating plateaus. Divergent actualization is possible both in structural and post-structural dynamics as an instantiation of bifurcating trajectories in the space of possibility, but in structural morphodynamics trajectories lead only to stable singularities.

The assemblage is namely the logic of conjunction of multiplicities in a never ending concatenation of and, and, and … . When we refer to heterogenetic dynamics we are thus dealing with the becoming of forms that are neither structures, on the one hand, nor sheer chaos, on the other. Between the dichotomy of chaotic dynamics, as the complete absence of forms, and stable symbolic structures, there is something else, which is what interests us. There is the possibility of a becoming of forms so rich that exceeds structures, changes laws, and recombines existing dynamics.

We wonder about the emergence of these forms and about the conditions of their emergence. We question the conditions of composition of the virtual in this dynamic deployments, that is the concept itself of differential.

1.3 The Weakening of Differential Constraints

Mathematics has long since set down the path towards weakening the concept of differential, starting from the revolution of non-Euclidean geometries put in place by Bernhard Riemann.

If Euclidean geometry was considered by Kant as an a priori and universal condition, Riemannian differential geometry was born with the aim of overcoming this notion by questioning the uniformity and isotropy of Euclidean space. This elaboration lead to the introduction of the notion of manifold, which is a more heterogeneous space that can be different from one point to another, since it is obtained via different dilations of the Euclidean space in the neighbourhood of each point. For example, in general relativity the geometry of space induces the dynamics of particles, but the space itself is not fixed, since the displacement of particles can in turn change the geometry of the space. The power of these new concepts introduced by Riemann for the construction of space-time did not escape Deleuze, who made Riemannian geometry a central tool for defining one of his main concept: the heterogeneity of a multiplicity. Starting from the 60s, however, a new process of weakening of differential constraints started in differential analysis and geometry. In the field of pure mathematics the early works of Franchi Lanconelli (1962) [5], and Stein (1976), introduced a more general notion of manifold, which no longer requires that the space be conceived as a local deformation of Euclidean space. A radical heterogeneity was thus introduced, which bore the possibility of constructing a space with different generators and dimensions for every point. This new conception of space was given the name of Sub-Riemannian geometry. In this new setting generators of the space change on a point-by-point basis, with the possibility that they don't commute, which in turn allows for the introduction of uncertainty in the same spirit of Heisenberg's celebrated uncertainty principle in the homonym group. The path towards a very heterogeneous and uncertain geometry was thus opened.

Nevertheless, it's worth noting that if the wave of sub-Riemannian geometry has been able to heterogenize geometric differential constraints, the dynamic behaviour of material flows was still understood to be homogeneous and uniform in space. The concept of heterogenesis as a heterogeneous becoming of both geometry and dynamics was not yet completely grasped.

1.4 Heterogenetic Morphodynamics

Today, however, we are dealing with the possibility of providing a renewed epistemic depth to the conceptual elaboration of Deleuzo-Guattarian heterogenesis. We intend to do this by both reconsidering the mathematical research of sub-Riemannian geometry, on the one hand, and by further developing a concept of heterogeneity capable of taking dynamics into account. To do so entails rejecting the accusations of superficiality and irrationality that have been issued against this paradigm. Likewise, it means demonstrating that heterogenesis entails material dynamics that are fundamental to both the life sciences and the human sciences and that are also crucial to the development of any political ecology.

If in A Thousand Plateaus, a multiplicity is a differential manifold and an assemblage is a combination of multiplicities, our aim in what follows is to *freely interpret the concept of multiplicity in terms of sub-Riemannian geometry and the concept of assemblage as a composition of them both in their geometric and dynamic counterparts*. Particularly dynamic heterogeneity can be expressed as the choice of a different operator at each point, which can induce a variety of dynamic behaviours that changes in space and time. To illustrate the point, we could think of a dynamic of diffusion, such as heat diffusion, in a certain point and a wave dynamic in a different point. Of course, the space of possibility that is opened by the assemblage of the two will overcome the two identitary behaviours towards a new hybrid composition. More generally, the virtual at stake in the becoming of forms is now a constellation of operators with different possible dynamics at every point.

We will turn our attention to sub-Riemannian geometries that allow degenerate metric introducing cuts and fragmented regions onto the manifold. In such a manner, fragile and temporary structures can be preserved by the construction of local spaces rather than being forced to be part of a global one. Moreover, the non-commutativity of fields generating sub-Riemannian spaces introduce uncertainty far beyond the celebrated Heisenberg uncertainty principle defined in the homonym group. Uncertain dynamics are thus deeply encouraged and can take place nearly anywhere.

The spatially and temporally varying definition of differential constraints at stake herein is quite far from the usual differential calculus of mathematical physics. In this latter, the distributions of operators are spatially and temporally homogeneous. Hence mathematical physics can be obtained as a special case when the same operator holds in every spatio-temporal point. With the composition of heterogenetic assemblages, however, the character of mathematical physics, in which space is a priori given, is completely reversed. Instead, operators are primary and serve to define both the dimensions and qualities of space: every new operator composed within the constellation completely redefines the entire space of the assemblage. In opposition to any understanding based upon universal laws, heterogenetic composition thus lays the conditions for an immanent morphogenesis, unfolding moment-by-moment by way of the assembling of singular concatenations.

To allow for the construction of assemblages, two temporal scales or axes must be present. The first is the axis of the actualization of differential constraints: the axis of Chronos common to mathematical physics. The second is the axis Deleuze calls Aion, along which takes place the recombination of differential constraints into new assemblages. On this axis, we find a true plasticity of the virtual - that is, the possibility to recombine genetic elements in order to create singular dynamics. Any specific recombination has to be thought of as an exploratory action, closer to a Dada performance than to a finalized process. Historicity is grounded in the possibility of memorizing configurations of differentials that have been successfully integrated, and using them as new genetic elements in future compositions.

Differential heterogenesis, then, is to be understood as the dynamics emerging from the composition of heterogeneous differentials that:

- Concatenate multiplicities, creating an immanent local space of possibilities.
- Introduce not only global but also fragmented and cut possibility spaces, creating autonomous zones and islands of the virtual.
- Extend the Heisenberg uncertainty principle to a variety of uncertainties given by any combination of non-commutative generators of space. Such a continuous recomposition of virtualities allows for the sudden emergence of coherent forms as well as ephemeral and rarefied dynamics in locally shared possibility spaces.

1.5 The Empirical Basins of Heterogenesis

Heterogenetic dynamics attain to different empirical basins. If the empirical basin of dynamic structuralism explored by René Thom and Jean Petitot is one of embryogenesis, that is, the set of dynamics at the base of the formation of biological bodies, with their symmetry breaking controlled by a parameter space, the main empirical example of post-structural dynamics is the brain. The brain is the body without organs par excellence, the body that, thanks to plasticity, changes its rules in a dynamic way and rebuilds itself continuously in a situated way. From here comes the necessity to think cerebral dynamics in the most heterogeneous possible way, taking into account the variety of neurogeometric architectures as well as the complex dynamics of neurotransmitters. Heterogeneity becomes even more radical if body-brain relations are taken into account and, more broadly, if cognition is considered in terms of enacted, embodied, embedded, extended processes.

On the other hand, we find post-structural dynamics in the life sciences when considering the evolution of living forms along a phylogenetic axis, on which generative elements are recombined. We thus find ourselves confronted with a double temporal axis: one pertaining to ontogenesis, wherein living forms are actualized, the other pertaining to the axis of phylogenesis, wherein generative constraints mutate. We can recognize here once again the two temporal axes of post-structural dynamics: They correspond to the axis of Khronos, where actualization takes place, and the one of Aion, where the recombination of differential constraints forms new assemblages and new configurations.

Such post-structural dynamics are also key elements of historical becomings. This can be seen, for example, from the perspective of micro-history, that is, histories understood by way of the dynamics of forms and the becoming of morphologies in the groove traced by Walter Benjamin. Micro-historical dynamics [3] are a laboratory for the morphology of multiplicity, in stark contrast to the forms of contemporary historiography that presents history as a progressive development of global phenomena that would characterize, in a uniform manner, the whole of a society, beginning with supporting structures and extending to symbolic and relational forms. This historiographic reduction is implemented both in space and in time: In space, the same phenomena would be uniformly present in the whole of the society, while in time, the same logics would unfold in epochal long ranges. To the contrary, it is important to reconsider and indeed reactualize the heterogeneity of forces and the variety of syncretic assemblages at the origin of historical dynamics [4]. Rather than a provisional quantitative model, heterogenesis is understood as a morphological device for developing a qualitative understanding of the generation of forms.

1.6 The Emergence of Semiosis

For this reason, heterogenesis is also at the centre of the question of the emergence of meaning in relation to the automatism of information processing. What is the difference between information processing and production of meaning? We can begin to consider this question by way of an example. For years we have studied and modelled cerebral processes in terms of information processing. It's true that this can now be accomplished with a less reductionist approach than was the case with the cybernetic cognitivism of the 60s. Today, contemporary technics of convolutional deep learning make it possible to construct AI architectures starting from databases pertaining to various types of phenomena (images, sounds, texts). But even with this refinement of technology, we are still dealing with information processing, albeit in highly sophisticated forms. The networks in question do nothing beyond reformatting statistics of data (stimuli or more elaborated data), remaining entirely dependent upon this data. What is experimentally observed in cerebral dynamics, however, is radically different from this. In this latter case, what is observed is that cerebral morphologies depend not only on external stimuli, but also on the presence of the situated body, that is both the cinematic-dynamic body with its mechanical constraints as well as the warm body with its large regulation systems linked to sexuality, nutritional circuits, feelings, and emotions. This has been the main research topic of Spinozist neuroscientist Antonio Damasio. The presence of the body modulates cerebral morphologies by means of mechanisms of reinforced learning in such a manner that morphologies reinforced by embodied feedback remain active. Cerebral circuits are then selected on the base of their major or minor meaningfulness for the situated body.

We rejoin here the Thomian theory of meaning, on the basis of which the French mathematician refounds his own physics of sense in a much less structural way than his previous catastrophe theory. The theory of meaning consists of the thesis that significant forms are constituted when bodily pregnancies take and modulate salient forms. This theory is then consistent with the approach of reinforcement learning, whose power goes far beyond the automatic response to Pavlovian stimuli. This has been shown by Patrizia Violi, for example, semiotician of the school of Umberto Eco, in a series of very interesting works that deal with the emergence of primary semiosis in the relationship between the mother and the newborn child. It is a question of semioses that, though pre-symbolic, are nonetheless trans-individual and social, irreducible to any disembodied processing of information.

In what follows, we will try to show that the saliences of the world and bodily pregnancies emerge from the same dynamic process in which two instances are just harmonic polarization of the heterogenetic flow. In other words, starting from the multitude of differential processes that constitute the morphogenetic power of nature, the actualization of these processes gives rise to the forms of the body and the forms of the world (if we want to see it in a Merleau-Pontyan perspective) and/or gives rise to a stratification of saliences and pregnancies (in the terms of Thom's semiotic system) or again, gives rise to layers of expression and content (if seen from a Deleuzian perspective). In this last case, the body that constitutes itself is already a multiple, infra-individual, and trans-individual social body.

The expressive space that has opened up in this way does not exclude the dimension of the sign. If capture devices such as Thomian catastrophes are present in the expressive space, the dynamics can bifurcate in such a manner as to bring along with it a semiolinguistic perspective.

We call semiogenesis the entire process of opening expressive space and installing within it, in a manner that is correlated to a plane of expression, the capture device of the signified.

1.7 Semiolinguistics

And so, it is through the prism of heterogenesis that we focus on some of the main questions of semiolinguistics—in particular, (i) those relating to the semiotic function, which is the essence of expressive phenomena and, (ii) those relating to semiogenesis, as a necessary "exit" from the expressive fact through its polarization as a sign. It is the constitution of expression/content substances and the fact of consubstantiality that are fundamentally questioned head-on. For this purpose, it will first be a matter of setting out a problematic framework. We will start by discussing the difficulties of conceptualizing the semiotic function in its most general modalities, namely in an epistemological light. As such, we recall that the categories of a "classical" epistemology, namely, those of form and substance, are an obstacle to the recognition of the principles and internal forms of the expressive phenomenon—it being understood that substance designates an actual and homogeneous diversity in which a form, which is the exclusive object of all knowledge, installs a relational network and therefore intelligibility. Correlatively, it is the relations of "exteriority" (partes extra partes), which characterize the order of empirical phenomena, which appear to be incompatible with the "interior" nature of the constitutive relations of expressive facts.

In order to overcome these two pitfalls, it will therefore be necessary to provide ourselves with a problematic apparatus that approaches the semiotic phenomena from prior to or "below" the articulation form/substance, and that, correlatively, gives an account of the "interiority" of the relations at the basis of the semiotic fact. This is precisely the contribution of the heterogenetic approach, which, as we will show, formally fulfils some of the main intuitions and conceptions of Merleau-Pontian phenomenology. In order to do so, it is necessary to mark out the problematic field. So, we first present the "truth" of the expressive fact and the resistances that it opposes to any conceptualization. Then, we will consider the different ways (operationalization, circumvention, externalization) that the "major" semiolinguistic theories have attempted to escape the aporias of the expressive fact. With all of this taken into consideration, the next step will be to introduce a problematic frame that responds to the need to approach the semiotic fact prior to the installation of substances (and together with substances, forms). To do this we will turn to the problematic frame of "solicitations" and "internal" relations, as developed by Merleau-Ponty, essentially in The Structure of Behaviour and Phenomenology of Perception.

The idea of "solicitations" refers to a pre-sensory halo (as the idea of differential refers to a pre-phenomenality), which is understood to be ante-substantial and to "question" and "interest" a vital principle, which latter responds to it by attributing to it a "corporeal existence" (i.e. a significant presence for the said vital principle) that in return, in a process of osmosis (interior relation) is validated by the rhythms of this halo. Now, it is precisely this dynamic of elaboration of the percepts as signifying presences that the heterogenetic approach accounts for—very precisely through the installation of harmonically correlated sub-substances in the emergence of a common empirical flow. To support this theoretical perspective, we take up, again following Merleau-Ponty, the question of first speech, and show how a theory of solicitations is required to go from the phoneme to the morpheme, i.e. from the differential fact to the expressive fact. However, once this theoretical device is accepted, certain difficulties arise. For the expressive fact encloses the perceiving or speaking subject in a "gangue" that absolutely conditions his field of possible actions (expressive and corporeal) and commands its executions. But the characteristic of semiotic systems, and more broadly of cultural systems, is that they constantly renew themselves by denying and surpassing themselves. It is therefore important to establish, "above" the heterogenetic device, a properly semiotic montage that takes such latitudes into account. This will be done by reinvesting previous works on the morphodynamics of the sign. These works demonstrate that the consubstantiality produced through the heterogenetic mechanism can be invested by morphodynamic processes that institute, according to links of determination, differential identities of expression and content, and correlatively introduce a dissymmetry between signifier and signified. Based on this asymmetry, which makes it possible to escape from semiolinguistic legality, the morphodynamics of the sign effectively account for the practical and creative freedom of the signifying apparatuses.

To summarize, the conception, argued along these lines, of a heterogenesis of substances and its necessary extension into a semiotic device, assumes that: (i) the encounter of a "bundle" of original solicitations is formally expressed as the elaboration of an assembly of differential operators with intensive value, (ii) the expressive space where expression and content substances are co-constituted is the result of a harmonic analysis of differential assemblages, and the substances of expression and content are configured for their own account as eigenvectors of this assembly, and (iii) morphodynamic processes invest these substances in order to instal a universe of signs and meaning, in other words a semiotic order.

1.8 Towards an Extended Imaginative Plane

Heterogenesis is the dynamics of an ecology of immanence, where the conditions of possibility of becoming are constantly changing in a manner fundamentally different from structural dynamics, which are based on the control of dynamics implemented as systems of gradient potentials in pregiven phase spaces, and whose effect is to categorize the control space by way of the instantiation within it of a network of boundaries. Heterogenesis, to the contrary, is an imaginative dynamics that can either generate completely new symbolic structures or remain a-symbolic, escaping any capturing and control device. In this sense it is an ecology of immanence. We thus propose that before any other attempt at ecological thinking, we must overcome the naive idea that imagination is a prerogative of symbolic elaboration and that aside from this exception there is only automatic stimulus-response. The ecological problem therefore deals, to a certain extent, with the problem of automatisms. At a dynamic level, one exits the automatism when one gains complete access to the axis of the historicity of processes. Following Deleuze, as we've already shown, the dynamics takes place on two temporal planes: the plane of Khronos, i.e. the one of the automatic actualization of differential constraints and on the plane of Aion, which comprises access to the past and the possibilities of the future. Both axes must be understood to be in continuous recomposition. The liberation from automatism lies in gaining access to this imaginative plane and in the ability to recombine intensive elements along this plane. It is the axis of phylogenesis in the evolution of species,[1] of the invention of the new in cognitive process, it is the plane of the uprising in social dynamics (uprising, not revolution, which is instead the structural concept of the transition from one stable system to another).

---

[1] Let's notice that also in ontogenesis effects of plasticity of the virtual and mutation can take place. Phylogenesis is just taken as the main example of mutation on the axis of Aion.

It's crucial to specify: these planes that we describe, those on which heterogenesis unfolds, are not a privilege of humans but open to an imaginative materialism which is extended to the animal, to the vegetable, to the inorganic... We are dealing with a generative materiality, capable of creating singularities at all scales and having to do with a vibrating flesh in continuous recombination. It is the multiplicity and diversity of the virtual that witnesses a continuous search for the new, a continuous re-imagination of the intensive, in opposition to a view of nature as a static system or as a depository of immutable laws. Nature is the process where "the real emerges as a dynamic, immanent multiplicity in a state of continuous variation, a metasystem far from equilibrium, rather than a combinatory manifestation or grammatical implementation of transcendent principles or rules, and as a differentiating relation, which is to say, as a heterogeneous disjunctive synthesis instead of a dialectical (horizontal) conjunction or hierarchical (vertical) totalization of contraries. ... So a new image of thought that is at once nomadological and multinaturalist" in the extent that "multinaturalism affirms not so much a variety of natures as the naturalness of variation—variation *as* nature". [9]: 105.

So, it's of crucial importance to abandon the reductionist perspective according to which the creation of sense would be tied exclusively to semiolinguistic aspects of human cultural production. Instead, it is necessary to be open up to much richer primary semiosis, an idea of meaningful forms generated by any encounter between salient forms in the world and bodily, affective pregnancies. This meeting between saliences and pregnancies gives rise to forms of primary signification, well before any emergence of signs as a system for communication, as shown very well by René Thom in a series of works that, as we've considered, establish a non-structuralist approach to the question of meaning. Both world's saliences and bodily pregnancies, furthermore, emerge from the same heterogenetic flow by way of multiple polarization.

The fact that these semioses are present in every heterogenetic becoming can offer a new manner of approaching the relationship between the human and the non-human. It is therefore a question of putting at the centre of our studies the conditions of production of sense that open up the possibility of creating planes of sensitive knowledge extended to technological, social, and ecological dimensions.

1.9 Organization of the Volume

The volume is organized as follows. In this chapter "Elements of morphodynamics" we recall some basics of morphogenetic becoming, starting from the idea of individuation in Gilbert Simondon, reconsidered explicitly in differential terms by Gilles Deleuze. Then, we briefly reprehend some principles of structural morphodynamics, as in the tradition of catastrophe theory, considering a particular kind of dynamics—i.e. gradient potential—that are at the base of the construction of structures. We will outline that structural dynamics, behind the richness and generative capability that it offers, in fact corresponds to a dynamics of control in a-priori given phase space, in which values are positional and differences are oppositional.

In the subsequent chapter, Multiplicity and Assemblages, the problem of the composition of differential fields, as it was faced by Deleuze and Guattari, is introduced. Here, the main object of study is the assemblage (agencement) as a composition of differential multiplicities. More specifically, we being by recalling the mathematical basis of heterogeneous multiplicity as a Riemannian manifold so as to introduce the concept of assemblage as a composition of manifolds. We refer to the main mathematical concepts presented in Difference and Repetition as well as in A Thousand Plateaus and attempt to clarify them in terms of the formalism originally employed by Bernhard Riemann and Albert Lautman. We choose to limit our focus to the mathematical aspects of these multifaceted philosophical texts; this allows us to progress towards our own mathematical construction.

The third chapter, "Differential Heterogenesis", inspired by these concepts of multiplicity and assemblage, develops a mathematical theory of differential becoming, wherein the composition of heterogeneous differentials—capable of generating new spaces of possibilities and new dynamics—is formalized. The introduction of manifolds based on sub-Riemannian geometry and the composition of heterogenous operators allows for the construction of heterogeneous dynamic assemblages, while the (partial) integration of dynamic constraints in these spaces instantiates a heterogenetic flow whose vibrational modes can be understood as plateaus that open onto spaces of expression. A concurrent dynamics of operators, spaces, and forms is then put into play to instantiate an immanent becoming.

In the chapter "Differential cognitive neuroscience", we consider the concept of differential heterogenesis from the point of view of cognitive neuroscience. In particular, we will consider neurogeometries and the neurodynamics of sensory cortices as well as the concatenation of cerebral micro-areas, which offer concrete examples of the assemblages in question. In fact, we will show that the phenomenal event—that is, perception and thinking—emerges precisely as the integration of forms from heterogeneous neural assemblages. We will show that eigenvectors of heterogeneous operators—that is, plateaus—correspond to mental objects in the most general sense and will provide some hints as to the manner in which this construction can be extended to 4E (embodied, embedded, extended, and enacted) cognitive sciences. The entire chapter is constructed in such a manner as to support the following point: if explicit mathematical models are possible for low-level perception, that is organized by principles of Gestalt laws and modal/amodal completion, such processes become more and more immanent and escape any claim of specific modelling when one moves to 4E perception and cognition. After such a transition, the only remaining possibility for grasping such dynamics is to individuate qualitative heterodynamics, which is precisely our objective.

In the chapter entitled "Semiogenesis" we focus on semiolinguistic phenomena in order to show how the heterogenetic perspective profoundly reconfigures the modalities and horizons of questioning about signs and about meaning. Indeed, the heterogenetic approach makes it possible to remove the main obstacle to semiolinguistic analysis, namely the enigmatic "consubstantiality" of the expressive fact, also called the "semiotic function". The key to this advance, which opens up the determination of phenomena hitherto belonging solely to the interpretative register (reflective judgement), consists in situating oneself, by way of heterogenesis, primitively "below" the substances (and their attendant forms) so as to account for their co-genesis. The empirical meaning of this heterogenesis and of the plane of virtualities prior to substances is given in the framework of an existential phenomenology (Merleau-Ponty), and specifically through the concept of "solicitation". Further, the interdependent substances that emerge from the heterogenetic process are understood to constitute the substrates (of expression and content) where, at a later stage and going beyond pure expressiveness by way of semiogenesis, morphodynamic operations will set up, in relation to one another and in a "control" relationship, differential identities of meaning (signifieds) and expression (signifiers).

Finally, a Chiusa will summarize the variety of heterogeneities we encountered over the course of the previous chapters and a section devoted to Plates of heterogeneous becomings will close the book.

References

1. Deleuze, G.: Difference and repetition. Columbia University Press, New York, 1968 (1994)
2. Deleuze, G., Guattari, F.: A thousand plateaus. In: Capitalism and Schizophrenia, vol. 2. University of Minnesota Press, Minneapolis (1980)
3. Ginzburg, C.: The cheese and the worms: The cosmos of a sixteenth century miller. Johns Hopkins University Press, Baltimore (First published in Italian as Il formaggio e i vermi, 1976) (1980)
4. Gribaudi, M.: Paris ville ouvrière, Une histoire occultée (1789–1848). La Découverte, Paris (2014)
5. Hörmander, L.: Hypoelliptic second order differential equations. Acta Mathematica **119**, 147–171 (1967)
6. Sarti, A., Citti, G., Piotrowski, D.: Differential heterogenesis and the emergence of semiotic function. Semiotica **230**, 1–34 (2019)
7. Simondon, G.: L'individu et sa genèse physico-biologique (l'individuation à la lumière des notions de forme et d'information). J. Millon, Paris, 1995 (1964)
8. Thom, R.: Structural stability and morphogenesis. University of Warwick, 1972 (1989)
9. Viveiros de Castro, E.: Cannibal metaphysics. Univocal, Minneapolis, 2010 (2014)

Chapter 2
Elements of Morphodynamics

2.1 Individuation

In his doctoral thesis of 1958 "L'individuation à la lumière des notions de forme et information" [28], Gilbert Simondon begins to elaborate the concept of individuation in terms of a passage from a pre-individual field to constituted forms. His thinking is deeply influenced by his mentor Maurice Merleau-Ponty and his thesis director George Canguilhem. In his idea of individuation, the becoming of forms is characterized by the transformation of a pre-individual field, comprised of intensive and genetic elements, to a plane of forms extended in space and time. His primary interest resides in the notion of individual and the way the individual is formed. By conceiving of individuation as a continuum and as a permanent process, Simondon intends to clarify that there is not a unity of identity but, to the contrary, that the individual is constantly *in fieri*, in the process of coming into being. Instead of a unity of identity, Simondon considers the individual as a transductive unit, meaning a process, in constant evolution, of the emergence of forms by way of contact.

Subsequently, the question of individuation is extended by Simondon: not only does it apply to the constitution of the human individual, but it is a universal morphological category, applicable to the generation of all forms, from physics to the life sciences, from psychology to social organization. Individuation, in short, becomes the centrepiece of any morphogenesis. As Deleuze outlines in his review [9] of Gilbert Simondon's thesis, individuation is everywhere: "Since the individual is 'placed' after individuation, the principle of individuation is 'placed' before the individuating operation, and consequently, above individuation itself; From this point on, individuation is 'placed' every-where; it is considered to be co-extensive with being, at least with concrete being (...). In truth, the individual can only be contemporaneous to his individuation, and individuation, contemporaneous to its principle: the principle must be truly genetic, not simply a principle of reflection".

In morphodynamical terms, however, Simondonian individuation has more than one interpretation and cannot be reduced to a specific dynamics. Gilbert Simondon's conception of individuation was of a process of differentiation—or, "répartition de

© Springer Nature Switzerland AG 2022

A. Sarti et al., *Differential Heterogenesis*, Lecture Notes in Morphogenesis,
https://doi.org/10.1007/978-3-030-97797-9_2

l'être en phases" [28]. Many of his readers [3, 21] have outlined that this differentiation, more specifically theorized by way of the term transduction, can be viewed as the dynamics of a metastable system allowing multiphase solutions. According to such an interpretation, the metastable system allows for multiple conditions of stability in such a way that the individuated field should be constituted by a plurality of phases. Phase transition is then the dynamic process to change from one phase to another in a manner analogous to thermodynamic processes. Other readers [25] have outlined the Simondonian interpretation of individuation in terms of quantum mechanics, where the plurality of states of a system is given by the spectral decomposition of the Hamiltonian operator. From such a perspective, a multiplicity of harmonic solutions is at stake and the composite state is given by the superposition of this multiplicity.

It's important to note that the two morphodynamic interpretations of individuation are not completely disjointed. The reason for this is that the possible states of a system after symmetry breaking are included in the eigenvectors of the spectral decomposition of the operator (see, for example, [5]). Of course, there are also differences: while in classical physics the actualization of a specific state "virtualizes the others" [21], in the quantum-mechanical case a wide hybridization of states is permitted. Rather than advocating for a specific interpretation to the exclusion of all others, we retain here their plurality, thus preserving the main message of Simondon's work, that is, that being is neither unitary nor fully individuated, but multiple and in perpetual individuation.

Another important point to emphasize with respect to Simondon's conception of individuation is that the pre-individual state from which the process is understood to begin is never completely actualized. Rather, after individuation, there is always a remainder that acts as a potentiality for subsequent individuations. Individuation, then, is always a partial integration, which offers a contrast with respect to the differential problems of mathematical physics, in which the differential is exhaustively integrated in emerging solutions. We will refind such an understanding of partial integration or integration with the remainder when viewing the Deleuzian plateaus as modes of vibration.[1]

2.1.1 *Differential Becoming*

In both Difference and Repetition [8] and in the preface to the publication of Simondon thesis [9], Deleuze proposes a concept of becoming that is largely based on the Simondonian idea of individuation. Deleuze retains the genetic and differential character of the concept of individuation, outlining an analogy that will be at the base

[1] The reader interested in deepening their understanding of the topic of individuation in Simondon philosophy can refer to the main works of Jean-Hugues Barthélémy [4] and Andrea Bardin [3]. For a discussion of the relation between Simondonian individuation and physical processes please refer to the reflection of Jean Petitot [21].

of his entire philosophy of difference: « A metastable system, essentially, entails the existence of a disparation (...). Therefore, it implies a fundamental difference, like a state of asymmetry. If it is nevertheless a system, it is only insofar as difference exists in it as potential energy, as a difference of potential distributed within certain limits. It seems to us that Simondon's perspective can be reconciled with a theory of intensive quantities, since each intensive quantity is a difference in itself. » ([9], 45).

Deleuze re-interprets Simondon's account of the passage from the pre-individual field to individuation as a passage from a virtual plane to an actual one, or in terms of a transformation progressing from the virtual plane to its actualization. Unlike Simondon, Deleuze characterizes this passage in a specifically mathematical way: he reconsiders the Leibnizian concept of differential and defines the virtual as a multiplicity of differential operators. So doing, he explicitly makes a link between individuation and differential theory, claiming that the intensive pre-individual field is to individuated forms as a distribution of differential operators is to their integrated solutions:

« The importance of Simondon's thesis is now apparent. He rigorously distinguishes singularity and individuality by discovering the preliminary condition of individuation. The metastable, defined as pre-individual being, is perfectly provisioned with singularities that correspond to the existence and distribution of potentials. (Could we not make the same claim in the theory of differential equations, in which the existence and distribution of "singularities" differ in kind from the "individual" form of the integral curves in their neighborhood'?) ». [9] pg. 44.

This distribution of operators is intensive and not perceivable, (in Merleau-Pontian terms, this distribution is a halo of « sollicitations », that is the ante-sensory stuff that a living body originally encounters, the sensorial qualities of which it must elaborate as Deleuze remarks in his Course at Paris 8 of 10/03/1981:

> These intensive quantities are expressed, defined, only by their distance to zero (...). Therefore, it is completely normal that if the entities are intensive quantities, they are expressed by ratios of differential, since the intensive quantity is inseparable from a definition in relation to zero. And that the differential ratio is precisely that. Everything becomes luminous!". (Our translation)[2]

and again,
« Difference is not phenomenon but the noumenon closest to the phenomenon (...). Every phenomenon refers to an inequality by which it is conditioned. Every diversity and every change refers to a difference which is its sufficient reason. Everything which happens and everything which appears is correlated with the order of differences: differences of level, temperature, pressure, tension, potential, differences of intensity » ([8]:22).

---

[2] "Ces quantités intensives s'expriment, se définissent, uniquement, par leur distance à zéro (...). Dès lors, c'est complètement normal que si les essences sont des quantités intensives, elles s'expriment dans des rapports différentiels, puisque la quantité intensive est inséparable d'une définition par rapport à zéro. Et que le rapport différentiel, c'est précisément ça. Tout devient lumineux !.

It's worth noting in particular that if Simondon considers the main level of his elaboration to reside in the definition he provides of the becoming of the individual, Deleuze is primarily interested in the emergence of thinking itself: "Differential calculus is the algebra of pure thought", ([8]:182) he writes in a passage of Difference and Repetition, where the main model of his philosophy of the difference is explicitly stated. The main project of the philosophy of Deleuze, as outlined by Anne Sauvagnargues [26], is then to set up a transcendental empiricism able to explain the genesis of thinking. In this framework, thinking emerges always by integration of a system of differential elements, a system of relations between genetic elements: "If Ideas are the differentials of thought, there is a differential calculus corresponding to each Idea, an alphabet of what it means to think. Differential calculus is not the unimaginative calculus of the utilitarian, the crude arithmetic calculus which subordinates thought to other things or to other ends, but the algebra of pure thought, the superior irony of problems themselves—the only calculus 'beyond good and evil'" ([8]:181). Even if differential calculus is just a mathematical tool, the differential becoming is considered by Deleuze to be a general dialectic that overcomes mathematics. Perceived forms, as well as mental forms of thinking, are nothing but the solution of a problem posed by the multiplicity of differential constraints that constitute the virtual. Analogously, problems that are physical, biological, sociological, or semiotic find their solutions in different disciplines by actualizing differentials in a proper manner. Mathematical calculus, then, is just the diagrammatic, abstract machine pointing to the concrete machine where the differential is implemented in a multiplicity of different modalities. In other words, the origin of any morphogenesis is differential.

2.2 Singularities

2.2.1 Poincaré Singularities

The concept of singularity understood as an attractor of a dynamics has been conceived in a series of important works of many mathematicians like Henry Poincaré [24], Aleksandr Lyapunov [16], Joseph P. LaSalle and Solomon Lefschetz[14], in works that stand today as pillars of classic studies in differential equations.

Poincaré effectively invented this new branch of mathematics—the "qualitative theory of differential equations"—showing that, even if an explicit solution of a differential equation cannot be provided, a study of the stability of the system can still be furnished. In both his doctoral thesis and in a series of memoirs under the title "On curves defined by differential equations" (1881–1882), he focused his attention on the behaviour of stationary points of the dynamics under perturbations of the solution. For reader's convenience we recall here qualitatively some basic notions of dynamical systems, inviting the interested reader to more formal and complete presentations as in [2].

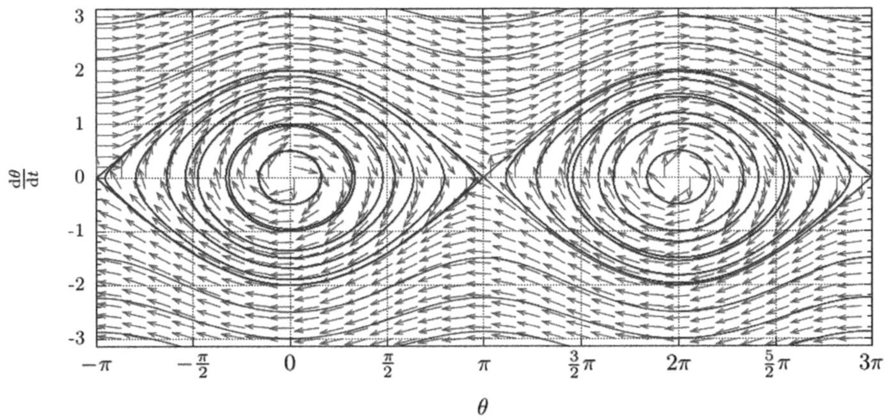

Fig. 2.1 Phase portrait of a pendulum in the phase space with coordinates the elongation angle θ and the angular velocity $\frac{d}{dt}\theta$

The evolution of a linear dynamical system is defined as

$$\frac{d}{dt}x(t) = Ax(t),$$

where x is the state of the system, $\frac{d}{dt}x(t)$ is its evolution in time and A is the evolution matrix.

Trajectories of the system are represented in the so-called phase space, that is the space having for axes the variables x and $\frac{d}{dt}x(t)$, that is position and velocity. A phase portrait is thus a geometric representation of the trajectories of the dynamical system in the phase space (see Fig. 2.1). Each set of initial conditions is represented by a different curve, or by the asymptotic solutions representing the behaviour of the system after a long period of time.

Qualitative theory of differential equations and dynamical systems studies the properties of these asymptotic solutions. This reveals information about the system such as the existence of attractors, repellors, or limit cycles for certain sets of parameter values. Two systems are topologically equivalent if their phase portraits represent the same qualitative dynamic behaviour. This relation of equivalence allows a classification of systems on the base of qualitative basins of their phase portrait (Fig. 2.2). The simplest kind of behaviour is exhibited by fixed points and by periodic orbits.

More generally, however, depending on the eigenvalues λ of the matrix A different kinds of behaviour give rise to different configurations of singular points. In the 2-dimensional case the classification is the following:

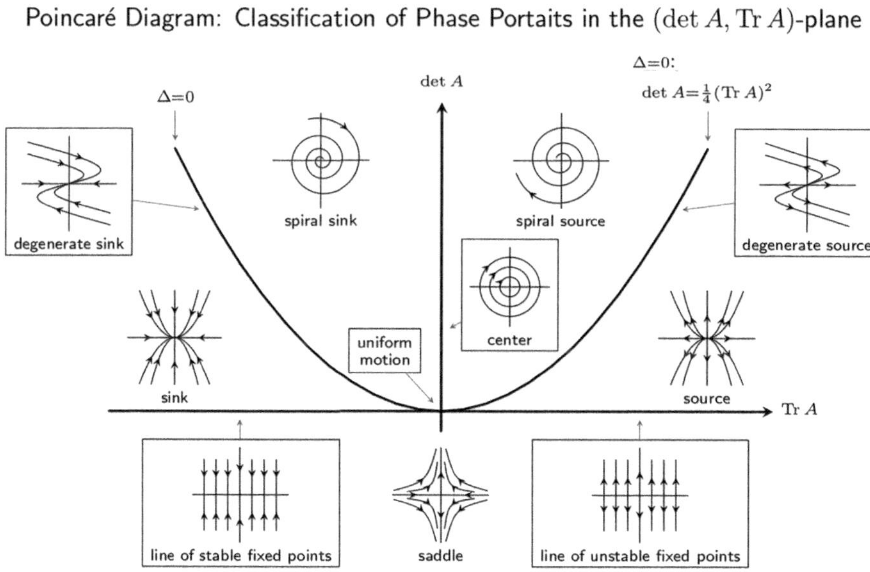

Fig. 2.2 Classification of critical points: Poincaré diagram in the plane (det$A$, Trace A) (form [10])

- Two distinct eigenvectors: Saddle, source, or sink.
  - Both negative eigenvalues: Sink.
  - Both positive eigenvalues: Source.
  - Mixed sign: Saddle.
- Complex eigenvalues: Centres and Spirals.
  - Real(λ) = 0 Centre (Purely periodic, closed curves).
  - Real(λ) > 0: Spiral Source.
  - Real(λ) < 0: Spiral Sink.
- One eigenvector: We have a degenerate source (positive λ) or sink (negative λ).

The extension of stability analysis to non-linear dynamical system is provided by the linearization of the dynamics near the equilibrium point.

Each of these points, as Poincaré has shown, induces an attractor basin, that is an area in the phase space inside which every trajectory is asymptotically attracted by the singular point. It means that every trajectory of the system, every integral curve, is moving towards a singular point. Intuitively, the attraction basin partitions the phase space in contiguous areas, each of which is related to a singular point. We will see that if all the attraction basins are governed by gradient potentials, as structuralists considered to be the case, the topological distribution of the qualitative types of those dynamics defines (through a network of boundaries) structural categories, and the entire dynamical system appears to be a categorizing device. To this end, the Catastrophe Theory proposed by René Thom provided the tools allowing for the passage from one stable basin to another, that is from one category to another, within the framework of the so-called structural morphodynamics.

2.3 Structural Morphodynamics

2.3.1 Degenerate Critical Points and Thom's Catastrophe Theory

Certain kinds of systems tend to minimize scalar energy, which means that the trajectories of the system follow the gradient of a smooth, well-defined potential V. Such systems are called gradient flows and are defined by trajectories that are a solution of:

$$\frac{d}{dt}x(t) = -\nabla V(x),$$

where $x(t)$ is the state variable, $\frac{d}{dt}x(t)$ is its variation in time and $\nabla V(x)$ is the gradient of the potential $V(x)$.

In general, these flows have attractors that are stable equilibria corresponding to the minima of the potential function. In correspondence of critical points the gradient of the potential function vanishes $\nabla V(x) = 0$ while generally the Hessian, i.e. the matrix of second derivatives, is not singular.

If for simplicity we consider V as a function of a single variable, the gradient is just the first derivative $\frac{dV}{dx}$ and the Hessian is simply the second derivative $\frac{d^2V}{dx^2}$. In this case, a non-degenerate critical point is a local maximum or a local minimum $\frac{dV}{dx} = 0$, depending on the sign of the second derivative, which is positive for a local minimum and negative for a local maximum. If also the second derivative is null $\frac{d^2V}{dx^2} = 0$, the critical point is called degenerate. In this case a small perturbation can change the sign of the second derivative and induce a loss of stability of the system.

To stabilize the system in these points we have to introduce a suitable control space $p = (p_1, p_2, \ldots, p_n)$ so that the system becomes

$$\frac{d}{dt}x(t) = -\nabla V(x, p),$$

where the potential V is now controlled by parameters p.

The behaviour of degenerate critical points for small variation of control parameters was the primary subject of study of René Thom's Catastrophe theory. Small changes of control parameters p change the shape of the potential $V(x, p)$ and can cause equilibria to appear or disappear, to change from attracting to repelling and vice

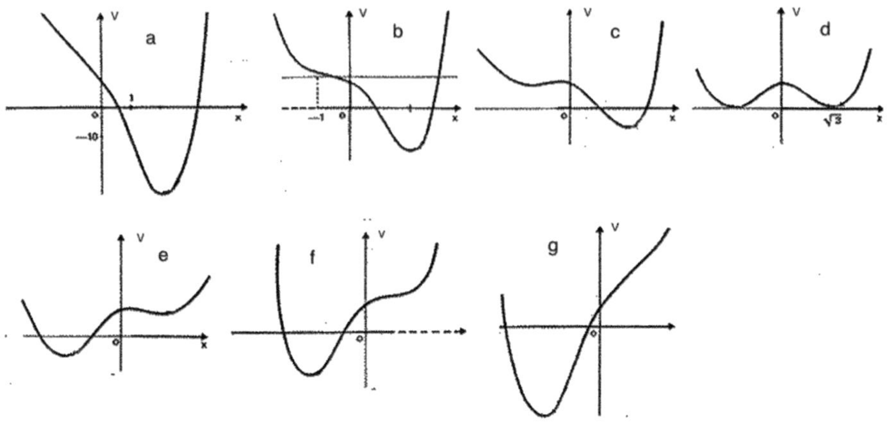

Fig. 2.3 The potential $V(x, p_1, p_2)$ of the catastrophe of the cusp by varying the control parameter p_1

versa, leading to large and sudden changes in the behaviour of the system. The space of state variables is called "internal space" while the space of control parameters is called "external space".

The points at which qualitative changes in behaviour occur are called catastrophe points. They correspond to germs where $V(x, p = 0)$. A smooth change of a parameter may result in drastic (discontinuous) changes in system behaviour.

The evolution of the system underlies the hypothesis that "external dynamics" adjusting the parameters is slow, while the "internal dynamics" expressed by the time evolution equation is fast. Thus, the systems will never be out of equilibrium and the evolution will be slow enough so that at each moment the initial state of equilibrium will be restored.

Near degenerate critical points, catastrophe theory can provide a canonical form for the potential, which depends only upon the control parameters. It also provides a classification of how stable equilibria change when parameters in the internal space are varied. It shows that stability is lost in correspondence with degenerate critical points and provides a way to re-stabilize the system by a change in parameter value. In this sense, Thom's theory is a theory of the stabilization of gradient flows.

To make this more clear, let's consider a concrete example, namely the catastrophe of the cusp. The state variable is one-dimensional x and there are 2 control parameters $p = p_1, p_2$. The potential has the form $V(x, p_1, p_2) = x^4/4 - p_1x^2/2 - p_2x$ (see Fig. 2.3).

By varying the parameters the potential takes different configurations with one minimum (a), two minima (c,d,e), the other minimum (g). The points (b) and (f) are bifurcation points wherein a new critical point appears. The configurations (c) and (e) show one minimum prevailing on the other while in (d) minima are equal.

By varying the value of paramaters, the set of equilibrium points constitutes the equilibrium manifold $\frac{\partial V}{\partial x} = 0$ that visually characterizes the catastrophe set

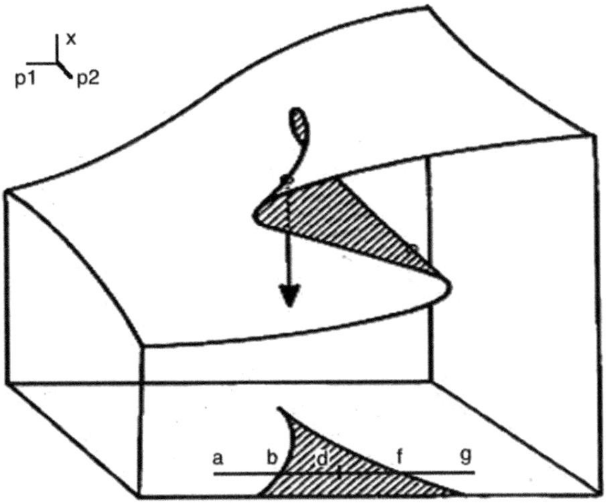

Fig. 2.4 Equilibrium manifold of the Cusp catastrophe. The bifurcation diagram is projected on plane (p_1, p_2). The points (a, c, d, e, g) correspond to configurations of the potential shown in the previous figure

Table 2.1 The potential of the seven elementary catastrophes

	Name	$V(x, p)$
1	Fold	$\frac{1}{3}x^3 + p_1x$
2	Cusp	$\frac{1}{4}x^4 + \frac{1}{2}p_1x^2 + p_2x$
3	Swallowtail	$\frac{1}{5}x^5 + \frac{1}{3}p_1x^3 + \frac{1}{2}p_2x^2 + p_3x$
4	Hyperbolic Umbilic	$x^3 + y^3 + p_1xy + p_2x + p_3y$
5	Elliptic Umbilic	$x^3 - 3xy^2 + p_1(x^2 + y^2) + p_2x + p_3y$
6	Buttefly	$\frac{1}{6}x^6 + \frac{1}{4}p_1x^4 + \frac{1}{3}p_2x^3 + \frac{1}{2}p_3x^2 + p_4x$
7	Parabolic Umbilic	$x^2y + y^4 + p_1x^2 + p_2y^2 + p_3x + p_4y$

(Fig. 2.3b). These configurations partition the parameter space (p_1, p_2) in three regions giving rise to the so-called bifurcation diagram (see Fig. 2.4).

One of the principal theorems of catastrophe theory stipulates that for at most four control parameters—thus the dimensions of space-time of our reality, there are only seven kinds of topologically distinct elementary catastrophes. In topology, forms that may appear as very different are in fact equivalent by continuous deformation.

We see in Table 2.1 the list of the seven elementary catastrophes defined in terms of the variables of state x, the control parameters $p = p_1, p_2, p_3, p_4$ and the potential $V(x, p)$. For every catastrophe it is thus possible to define an equilibrium manifold as $\frac{\partial V}{\partial x} = 0$ and its bifurcation set. Catastrophe theory can accordingly be considered as a bifurcation theory for gradient flows.

The fundamental theorem of catastrophe theory, in dimension 3 (three external parameters), states the existence of five sets of elementary catastrophes. Each of

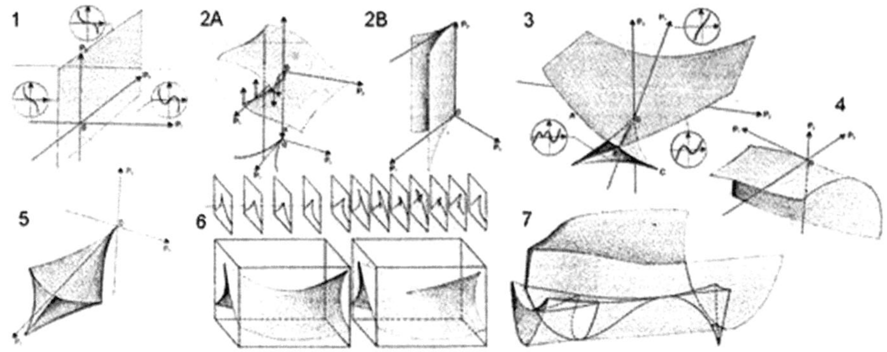

Fig. 2.5 Seven elementary catastrophes represented by the respective equilibrium manifold (from [11]). Each equilibrium manifold partitions the control space in regions with specific configurations of the potential that are shown inside the circles

them is associated with a system described by a potential dependent on one [(1), (2), (3)] or two internal variables [(4), (5)]. The extension to four parameters involves the introduction of two new forms ((6), (7)). Their representation becomes much more difficult, as we pass into dimension 4. It should be noted that the form of these sets is familiar to us and that their names are chosen to underline their resemblance to common objects (see Fig. 2.5).

2.3.2 *Dynamizing Saussure, Greimas, and Lévi-Strauss*

Since each equilibrium manifold partitions the control space into distinct regions, it becomes the perfect dispositive for categorization and the central device for defining "structures". Unlike formal structuralism (e.g. [13]) which produces its analyses at a logico-algebraic level, "classical" structuralism is based on topological and dynamical intuitions: In classical structuralism, a structure is defined as a system of oppositions categorizing some domain. Thom's Theory of Catastrophes [29] provides the morphological support necessary to generate this system of oppositions in such a manner as to open onto what is called "structural morphodynamics". Thom's theory, as we have seen, is essentially a theory of the stabilization of gradient flow, i.e. a theory of control that would allow to move a system from one stable state to another. A stable state is represented by the position of a point in the space of control parameters and the change of state is a path in the control space. For Thom the passage from one stable point to another is immediate and there is no interest in transitory dynamics. A catastrophe is thus a partition of the control space capable of categorizing different regimes of functionality of the system. For its essential "positional" character, a catastrophe is thus the naturalization of what structuralism defines as a "structure":

> The elements of a structure have ... nothing other than a sense: a sense which is necessarily and uniquely "positional." ... It is not a question of a location in a real spatial expanse, nor of sites in imaginary extensions, but rather of places and sites in a properly structural space, that is, a topological space. Space is what is structural, but an unextended, preextensive space, pure spatium constituted bit by bit as an order of proximity, in which the notion of proximity first of all has precisely an ordinal sense and not a signification in extension. ([7]: 103).

The first example proposed by Deleuze as a structure concerns, in full agreement with Thom's approach, the development of biological forms in embryogenesis:

> Take genetic biology: the genes are part of a structure to the extent that they are inseparable from "loci," sites capable of changing their relation within the chromosome. In short, places in a purely structural space are primary in relation to the things and real beings which come to occupy them, primary also in relation to the always somewhat imaginary roles and events which necessarily appear when they are occupied. [7].

The ambitions of structuralism, however, extend far beyond biology, being applied to trans-individual organizations like economy in Althusser's "Reading Capital", [1]: "And when Althusser speaks of economic structure, he specifies that the true 'subjects' there are not those who come to occupy the places, i.e. concrete individuals or real human beings—no more than the true objects are the roles that they fulfil and the events that are produced. Rather, these 'subjects' are above all the places in a topological and structural space defined by relations of production". [7].

And again, evoking Claude Lévi-Strauss's

> The Elementary Structures of Kinship Relations" Deleuze again outlines: "Father, mother, etc., are first of all sites in a structure; and if we are mortal, it is by moving into the line, by coming to a particular site, marked in the structure following this topological order of proximities (even when we do so ahead of our turn). [7].

Meanwhile, the word "structuralism" itself, it's worth recalling, originates in Ferdinand de Saussure's famous Cours de linguistique générale [6], wherein the Swiss linguist proposes to consider language as a system in which every element (qualified as a « value ») has to be defined by way of and at the point of intersection of two kinds of relationships: the (external) relations of equivalence (also called exchange) and the (internal) relation of opposition. This set of relations is understood to constitute the very "structure" of language.

To naturalize these systems of oppositions within the domain of morphodynamics, Jean Petitot proposed in the 70s to use Thom's Catastrophe Theory. The relation between catastrophe theory and structural semiotics has been widely studied in Petitot's principal work Morphogenesis of Meaning [18]. It was then further developed by Per Aage Brandt (P-A. Brandt, 1986) and additional linguistic applications have been outlined by Wolfgang Wildgen [30].

Before arriving at a morphodynamical solution, Petitot problematizes the relationship between universal categories and specific empirical domains, in this case semio-narrative structures, finding two possible responses:

(1) The first, logicist-formalist, favoured by Carnap, Hjelmslev ... is that of axiomatization. But, insofar as the categorial indefinables have no proper object, it only leads to a general logic of relations which is essentially trivial on the mathematical level (cf. the logical semiotic square to which we will return). One might think that it was partly supported by Greimas, but this is not so clear since he has often expressed his "unease" with regard to formal logic and axiomatization. For him, the term "logic" referred rather to the idea of a "linguistic calculation".
(2) The second, much more demanding and which I favoured from the start, is the transcendental response. It is inspired by what Kant did in his First Metaphysical Principles of a Science of Nature about Newtonian Mechanics. Mechanics concerns a specific primitive core phenomenon (Husserl would say a "regional object"): that of the spatio-temporal movements of material bodies." [20].[3]

In Petitot's structural morphodynamics, categorial equivalences and oppositions are implemented by means of control of stable attractor basins, where a category is selected by a certain configuration of control parameters in opposition to another category. Catastrophe Theory thus acts like a categorization device with respect to the states of a system. Owing precisely to its positional nature for its ability to implement oppositive categorization, it provides the basis for structural modelling, in the sense of Saussure, Lévi-Strauss, Jakobson, Greimas.

In fact, for Saussure there is no possibility of having a single sign alone, since signs define themselves reciprocally by way of relations of opposition. At least two signs are always needed. A morphodynamical model of the Saussurian sign has been provided by D. Piotrowski in [23], by means of the catastrophe of cusp (in its simplest form), that is the catastrophe which synthesizes the situations of conflict and of bifurcation. In this case, the two stable basins of the potential categorize the substance of content of the sign (Fig. 2.6). Analogously, for Greimas deep structures of signification are modelled by a semiotic square, which offers a sophisticated system of contrary, contradictory and implication relations [12]. The semiotic square has often been considered as the set of relations underlying analogy; and with respect to the simple opposition of the Saussurean sign in its basic form (see above), it's worth highlighting that the oppositive relations have been doubled, with the result that the possible stable states are now four. A morphodynamical model of the semiotic square can thus be represented by a double cusp catastrophe with germ $V(x, 0) = x^4 + y^4$. The study of this catastrophe is very complex and we recommend that the interested

[3] (1) La première, logiciste-formaliste, favorisée par Carnap, Hjelmslev ... est celle de l'axiomatisation. Mais, dans la mesure où les indéfinissables catégoriaux sont sans objet propre, elle n'aboutit qu'à une logique générale des relations qui est essentiellement triviale sur le plan mathématique (cf. le carré sémiotique logique sur lequel nous allons revenir). On pourrait penser qu'elle a été en partie soutenue par Greimas mais ce n'est pas si clair car celui-ci a souvent fait part de son « malaise » relativement à la logique formelle et à l'axiomatisation. Chez lui, le terme « logique » renvoyait plutôt à l'idée d'un « calcul linguistique ». (2) La seconde, beaucoup plus exigeante et que j'ai privilégiée dès le début, est la réponse transcendantale. Elle s'inspire de ce que Kant a fait dans ses Premiers principes métaphysiques d'une science de la nature à propos de la Mécanique newtonienne. La Mécanique concerne un phénomène noyau (Husserl dirait un « objet régional ») primitif spécifique: celui des mouvements spatio-temporels des corps matériels.

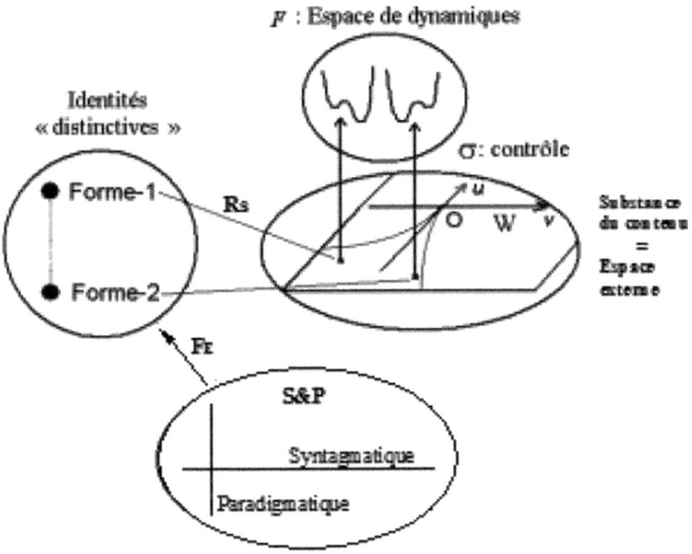

Fig. 2.6 Morphodynamics of the Saussurian sign with the cusp catastrophe [23]

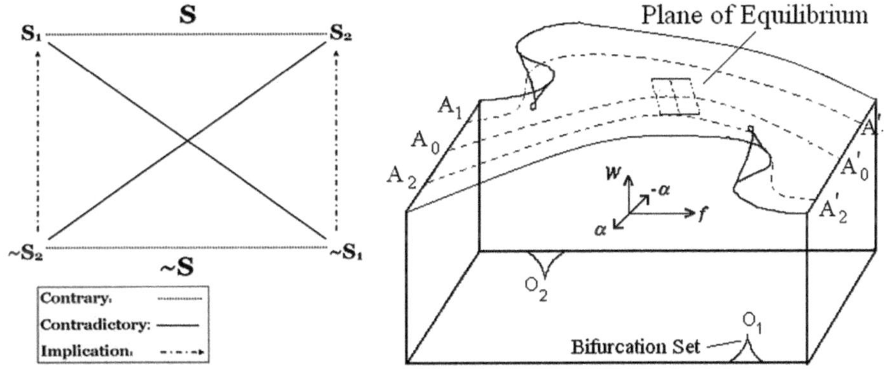

Fig. 2.7 Greimas semiotic square (left) and its morphodynamics with the double cusp catastrophe (right). The equilibrium manifold is shown as well as the two bifurcation sets B1 and B2

reader refers to specialist literature [31]. For our current purposes, it will suffice to outline that its bifurcation set implements four stable attractor basins (Fig. 2.7), which categorize the control space of the semiotic square. Alternative models of the semiotic square have been provided by [22].

A further doubling of opposition relations is necessary to represent the morphogenesis of the myth as it was structurally defined by Claude Lévi-Strauss in the famous article "The Structural Study of Myth" [15]. The canonical formula of myth proposed therein can be geometrically modelled in the group of Klein as proposed Lévi-Strauss himself and Lucian Scubla in [27] or in the group of quaternions following [17]. In this last structure, three pairs of oppositions are presently obtaining a partition of the internal state space in eight stable basins. For a debate about the morphodynamics of myths, see, for example, [19] and [27].

The simple structures we have recalled here are essentially elementary combinations of oppositions, the fundamental thesis of structuralism being, we repeat, that the positional difference is ontologically first in relation to identity. Elementary combinations of oppositions are then morphogenetically implemented by combinations of cusp catastrophes with different germs.

We recall here once more the importance of the contribution of René Thom in formalizing the passage from any stable state to another stable state by means of Catastrophe theory as well as the contribution of Jean Petitot in expanding the morphodynamic approach to structural semiotics by modelling elementary structures of the signification of Greimas (actantial syntax à la Propp and semantic paradigms) and the canonical formula of myth of Lévi-Strauss.

At the same time, we wish to emphasize that structural morphodynamics is based on a very peculiar kind of dynamic becoming, that is gradient flows, according to which flows lead necessarily to stabilized solutions. An abundance of different dynamics can, however, be envisaged, dynamics that would not necessarily give rise to stabilized singularities. Moreover, structural morphodynamics deals with bifurcations and catastrophes inside a given phase space while a variety of interesting phenomena, like imaginative events and living mutations, find their focus in the continuous composition of the space of possibilities that is not given a priori. To construct these latter kinds of dynamics, in contrast to the structuralist approach, can serve to introduce, from a geometrical as well as a dynamical perspective, the concepts of multiplicity and assemblage.

References

1. Althusser, L.: Reading Capital, trans. Ben Brewster. New Left Books, London, 1970 (1965)
2. Arnold, V.I.: Mathematical Methods of Classical Mechanics. Springer (1982)
3. Bardin, A.: Epistemology and Political Philosophy in Gilbert Simondon. Springer Editions (2015)
4. Barthélémy, J.-H.: Penser l'individuation: Simondon et la philosophie de la nature. Paris, L'Harmattan (2005)
5. Citti, G., Sarti, A.: A Gauge field model of modal completion. J. Math. Imaging Vis. **52**(2), 267–284 (2015)
6. de Saussure, F.: Course in general linguistics (W. Baskin, Trans.), vol. 36. Philosophical Library, New York, 1959 (1916)
7. Deleuze, G.: A qua reconnait-on le structuralisme? In: Châtelet, F. (ed.) Histoire de la philosophie VIII, Verviers Marabout (1973)
8. Deleuze, G.: Difference and Repetition. Columbia University Press, New York, 1968 (1994)
9. Deleuze, G.: Review of Gilbert Simondon's "L'individu et sa genèse physico-biologique", Pli, pp. 43–49. (This review article appeared originally in Revue Philosophique de la France et de l'Etranger 156 (1966)), 2001 (1966)
10. Egwald: (2019). http://www.egwald.ca/linearalgebra/
11. Ekeland, I.: La recherche, No. 81 **8**, 745–754 (1977)

12. Greimas, A.-J.: Structural Semantics. University of Nebraska Press, 1966 (1983)
13. Hjelmslev, L.: Prolegomena to a Theory of Language, F. J. Whitfield (trans.). University of Wisconsin Press, Madison, 1969 (1943)
14. Lefschetz, S., LaSalle, J.P.: Stability by Liapunov's Direct Method with Applications. Academic Press, New York (1961)
15. Lévi-Strauss, C.: The structural study of myth. J. Am. Folk. **78**(270), 428–444 (1955)
16. Lyapunov, A.M.: The General Problem of the Stability of Motion. Kharkov Mathematical Society, Kharkov (1892)
17. Morava, J.: On the canonical formula of C.Lévi-Strauss (2003). https://arxiv.org/abs/math/0306174
18. Petitot, J.: In: Brandt, P.-A., Wildgen, W. (eds.) Morphogenesis of Meaning. Peter Lang, Bern, 2004 (1985)
19. Petitot, J.: Approche morphodynamique de la formule canonique du mythe. L'Homme **106–107**(2–3), 24–50 (1988)
20. Petitot, J.: Mémoires et parcours sémiotiques du côté de Greimas, Actes Sémiotiques, n.120 (2017)
21. Petitot, J.: Physique du sens: de la théorie des singularités aux structures sémio-narratives, éditions du CNRS. , Paris (1992)
22. Petitot, J.: Topologie du carré sémiotique , Études littéraires (1977)
23. Piotrowski, D.: Morphogenesis of the Sign. Springer, Berlin (2017)
24. Poincaré, H.: Mémoire sur les courbes définies par une équation différentielle, (I)Journal de mathématiques pures et appliquées 3esérie, tome **7**, 375–422 (1881)
25. Sarti, A., Piotrowski, D.: Individuation and semiogenesis: an interplay between geometric harmonics and structural morphodynamics. In: Sarti, A., Montanari, F., Galofaro, F. (eds.) Morphogenesis and Individuation, pp. 49–73. Springer, Berlin (2015)
26. Sauvagnargues, A.: Deleuze, l'empirisme transcendantal. PUF, Paris (2008)
27. Scubla, L.: Hesiod, the three functions, and the canonical formula of myth. In: Maranda, P. (ed.) The Double Twist: From Ethnography to Morphodynamics, pp. 123–155. University of Toronto Press (2001)
28. Simondon, G.: L'individuation à la lumière des notions de forme et d'information. J. Millon , Paris, 2015 (1964)
29. Thom, R.: 1972. Structural Stability and Morphogenesis, University of Warwick (1989)
30. Wildgen, W.: Catastrophe Theoretic Semantics. An Elaboration and Application of René Thom's Theory. Amsterdam, Benjamins (1982)
31. Zeeman, C.: Catastrophe Theory: Selected Papers, 1972–1977. Addison-Wesley, London, Amsterdam, Ontario, Sydney, Tokyo (1977)

Chapter 3
Multiplicity and Assemblages

3.1 Multiplicity

For the Deleuze of "Difference and repetition", whose interest resided in the virtual conditions of the becoming of forms, the process of individuation must be conceived as a transformation from one multiplicity to another multiplicity: "Thus each individual is an infinite multiplicity, and the whole of Nature is a multiplicity of perfectly individuated multiplicities" ([6]: 254).

This definition of becoming is very different from the idea of morphogenesis proposed by the Gestalt school. Whereas this latter explains the emerging of forms as the articulation between a figure and a background, with unity of the two understood as the distinctive feature of the gestalt, the Deleuzian becoming is always plural, both at the level of the virtual plane and in its actualization. If for the Gestalt school a form is an organic object as in the tradition of German romanticism, Deleuzian becoming is a baroque variety of heterogeneities. Without a doubt, Deleuzian individuation is closer to Simondonian one, for whom the emergence of a form is inseparable from a plurality of pre-individual elements that are never fully actualized. This transformation from plurality to plurality becomes particularly important when multiplicities are put to work in order to explain the creation of concepts. Already in Difference and Repetition, Deleuze defines Ideas as the virtual of any production of thinking, corresponding to intensive multiplicities which are always primary with respect to any unity:

> Ideas are multiplicities: multiplicity must not designate a combination of the many and the one, but rather an organisation belonging to the many as such, which has no need whatsoever of unity in order to form a system." ([6] (1968): 182). And again: "This is not surprising, since becoming and multiplicity are the same thing. A multiplicity is defined not by its elements, nor by a center of unification or comprehension. It is defined by the number of dimensions it has; it is not divisible, it cannot lose or gain a dimension without changing its nature ([6] (1968): 249). At the same time, every multiplicity is already a heterogeneous composition of elements in continuous transformation since "each multiplicity is already composed of heterogeneousRiel terms in symbiosis, and that a multiplicity is continually transforming itself into a string of other multiplicities ([6] (1968): 249).

© Springer Nature Switzerland AG 2022

A. Sarti et al., *Differential Heterogenesis*, Lecture Notes in Morphogenesis,
https://doi.org/10.1007/978-3-030-97797-9_3

Three conditions pertaining to any multiplicity are thus made explicit: (1) virtuality, or the non-phenomenal character of the multiplicity (2) juxtaposition, or the principle of construction by the inside (3) differentiability, or the genetic feature of the multiplicity. These three conditions are elaborated as follows:

> (1) the elements of the multiplicity must have neither sensible form nor conceptual signification, nor, therefore, any assignable function. They are not even actually existent, but inseparable from a potential or a virtuality. In this sense they imply no prior identity, no positing of a something that could be called one or the same. To the contrary, their indetermination renders possible the manifestation of difference freed from all subordination.
>
> (2) These elements must in effect be determined, but reciprocally, by reciprocal relations which allow no independence whatsoever to subsist. Such relations are precisely non-localisable ideal connections, whether they characterise the multiplicity globally or proceed by the juxtaposition of neighbouring regions. In all cases the multiplicity is intrinsically defined, without external reference or recourse to a uniform space in which it would be submerged. Spatio-temporal relations no doubt retain multiplicity, but lose interiority; concepts of the understanding retain interiority, but lose multiplicity, which they replace by the identity of an 'I think' or something thought. Internal multiplicity, by contrast, is characteristic of the Idea alone.
>
> (3) A multiple ideal connection, a differential relation, must be actualised in diverse spatio-temporal relationships, at the same time as its elements are actually incarnated in a variety of terms and forms. ([6] (1968): 182-183).

These conditions establish an unequivocal link between the concept of multiplicity and the one of differential manifold as introduced in Riemannian geometry, a link that is doubtless deserving of further investigation.

3.2 Riemannian Geometry

3.2.1 Manifolds

Riemannian differential geometry was born with the aim of overcoming the notion of Euclidean space as a uniform and isotropic space. The power of the new concept introduced by Riemann for the construction of space-time does not escape Deleuze. Rather, Riemannian geometry becomes a central tool in his account of the genesis of ideas. The concept of Riemannian manifold becomes the main reference at stake in the definition of multiplicity. The french term "multiplicité" is namely the translation of the german "mannigfaltigkeit", which corresponds to the english word "manifold" and, more importantly, is used by Riemann to define his space:

> Ideas are multiplicities: every idea is a multiplicity or a manifold. In this Riemannian usage of the word 'multiplicity' (taken up by Husserl, and again by Bergson) the utmost importance must be attached to the substantive form: multiplicity must not designate a combination of the many and the one, but rather an organisation belonging to the many as such, which has no need whatsoever of unity in order to form a system. ([6] (1968): 182).

The main property of a manifold is that it is defined locally as a curved version of a Euclidean space. As such it has the same dimension as the Euclidean space at every point. The intuitive idea of a manifold is that it can be a curve if it has dimension one, a surface if it has dimension two, and so on—but in fact it can have any dimension whatsoever. Higher dimensional surfaces code more complex geometrical objects. Projection in each dimension can describe a single aspect of the surface. This is why dimensionality is explicitly quoted by Deleuze, to indicate the complexity of a multiplicity.

> A multiplicity ... is defined by the number of dimensions it has; it is not divisible, it cannot lose or gain a dimension without changing its nature. ([6] (1968): 182).

In what follows, we will always represent graphically only 2D surfaces, which can be easily visualized, but one must keep in mind that manifolds are multidimensional.

3.2.2 *Charts*

Precisely the manifold is locally a deformation of a piece of Euclidean space that can be mapped by a chart. In the following, the subset of the Euclidean space R^n (where n is the dimension of the space) will be denoted U_α and the chart is defined as an invertible function

$$U_\alpha \subset R^n, \quad \phi_\alpha : U_\alpha \to M, \tag{3.1}$$

where M denotes the manifold. The invertibility condition ensures that $\phi_\alpha(U_\alpha)$ is a deformed version of U_α without cuts (see Fig. 3.1, where the white grid on R^2 is mapped to a curved grid on the manifold M). But to map a manifold a single chart cannot be sufficient.

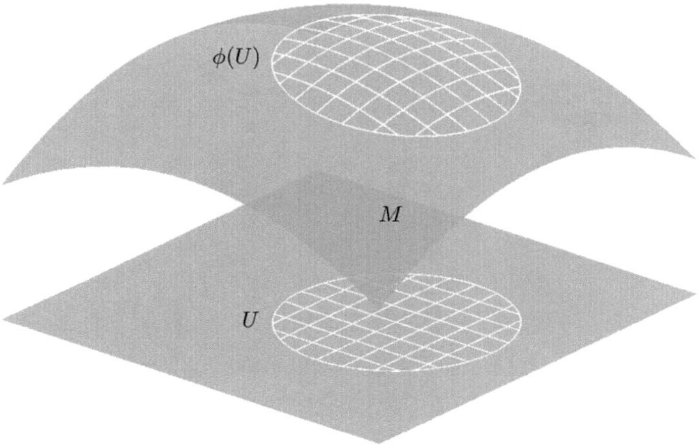

Fig. 3.1 Locally a manifold M is a piece of curved space, obtained via a deformation of a ball of Euclidean space. The function $\phi : U \to M$ defines a local chart

3.2.3 Atlas and Smooth Manifolds

A manifold overcomes the dichotomy between the multiple and the one because it is a unique geometrical object and a union of local charts:

> Multiplicity remains completely indifferent to the traditional problems of the multiple and the one, and above all to the problem of a subject who would think through this multiplicity, give it conditions, account for its origins, and so on. There is neither one nor multiple, which would at all events entail having recourse to a consciousness that would be regulated by the one and developed by the other ([6] (1986), 14).

Indeed a manifold is defined in terms of a set of charts, also called an atlas. For the sake of example, let us consider a sphere (Fig. 3.2). If we try to cover it with a unique

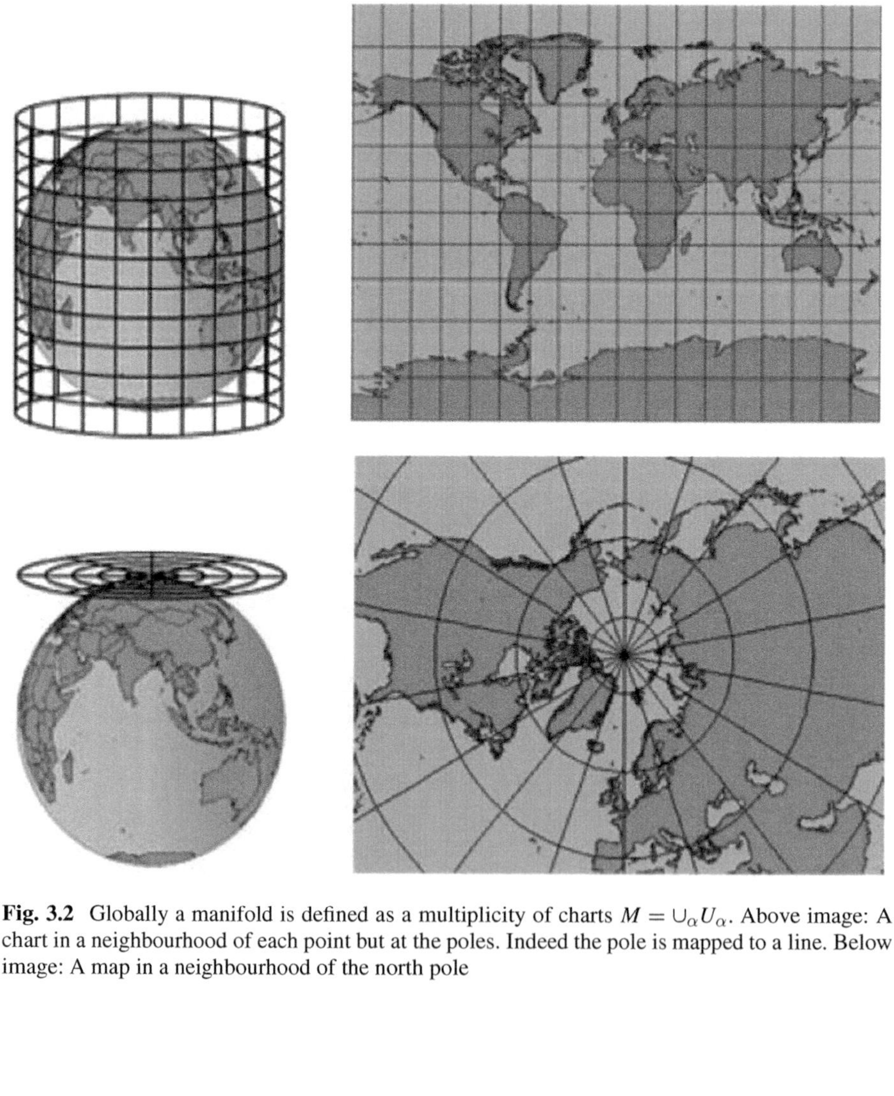

Fig. 3.2 Globally a manifold is defined as a multiplicity of charts $M = \cup_\alpha U_\alpha$. Above image: A chart in a neighbourhood of each point but at the poles. Indeed the pole is mapped to a line. Below image: A map in a neighbourhood of the north pole

chart, the pole becomes a line. The collapsing of a point in a line is not compatible with the definition of manifold. We need at least two charts to cover the sphere: one around each pole. We can, however, just as well consider an atlas composed of more than two charts. It is necessary to have more than one chart to describe the sphere, but none of the additional charts are necessary to the description.

Indeed an atlas can be formally defined as a collection of charts, each one is defined as in (3.1) as a deformation ϕ_α acting on a set U_α:

$$M = \cup_\alpha U_\alpha, \quad \phi_\alpha : U_\alpha \to M. \tag{3.2}$$

This is a first level of multiplicity. The discreet multiplicity of charts.

Now we must say what is the way to glue the charts to compose an atlas. It is clear that it is not enough to show two charts to understand how the atlas is structured, but it is necessary to show how they must be glued.

More specifically, if two charts ϕ_α and ϕ_β overlap, they provide two different deformations of the same region. In Fig. 3.3 we represented two charts, and for example, Greenland belongs to the intersection of their codomain $\phi_\alpha(U_\alpha) \cap \phi_\beta(U_\beta)$. As a result U_α and U_β contain two deformed versions of Greenland, and there is a smooth map between these two representations. Then to have a gluing of charts we formally require that the mapping

$$\phi_\beta^{-1} \circ \phi_\alpha : U_\alpha \to U_\beta \tag{3.3}$$

is smooth.

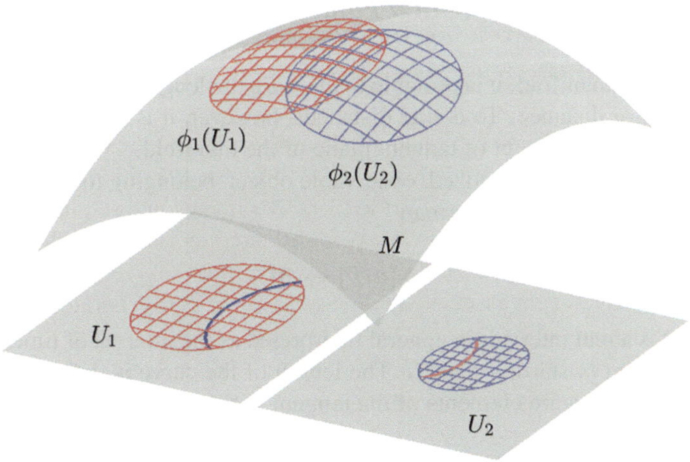

Fig. 3.3 The gluing of a chart to the next one is not defined a-priori and can be done in endless ways. This characterizes smooth manifolds as opposed to striated ones

This invertibility condition describes the deformation operated by the composition of the two charts and constitutes the condition of gluing of charts.

Multiplicity is thus a way to re-problematize the space by not assuming that it is already given, but rather constituting it piece-by-piece by proximity. In fact, a multiplicity is an amorphous collection of pieces that are just juxtaposed to one another. It is above all this condition, i.e. that there is an infinite number of ways of gluing maps that leads Deleuze and Guattari to the definition of smooth space:

> Smooth space is precisely the space of the smallest deviation: therefore it has no homogeneity, except between infinitely proximate points, and the linking of proximities is effected independently of any determined path. It is a space of contact, of small tactile or manual actions of contact, rather than a visual space like Euclid's striated space. Smooth space is a field without conduits or channels. A field, a heterogeneous smooth space, is wedded to a very particular type of multiplicity: nonmetric, acentered, rhizomatic multiplicities that occupy space without "counting" it. ([3] (1980): 371).

A smooth Riemannian space, then, is a heterogeneous space, "an amorphous collection of juxtaposed pieces that can be joined together in an infinite number of ways: we see that patchwork is literally a Riemannian space, or vice versa" ([3] (1980): 476).

The smooth space (espace lisse) is a *differential manifold* where we can only glue charts; we can define curves in such a space, but we do not yet have a notion of the length of curves, nor any metric distance. Hereafter, we will introduce these missing elements following the Riemann construction, so as to allow us to define, in addition to a smooth space, a striated space.

3.2.4 The Tangent Plane at a Point

In a Riemannian manifold, it is possible to measure the length of curves by introducing the notion of distance. To define distances, however, it is necessary to preliminarily introduce the concept of tangent plane of the manifold.

Indeed, a curve is an actualized, observable object, belonging to the manifold M. It is simply defined as a smooth map

$$\gamma : [a, b] \to M,$$

where $[a, b]$ is a real interval, in general interpreted as the interval of time necessary to reach the point b, starting from a. The length of the curve is defined in terms of its derivatives, which are elements of the tangent to the manifold, that is, the virtual of the manifold.

There is a tangent plane for every point of the manifold M: fixing a point p_0 in the manifold M the tangent to M in p_0 consists of the whole set of all the first derivatives of curves on the manifold passing through the point p_0.

$$T_{p_0}(M) = \{\gamma'(t) : \gamma \text{ is a smooth curve on } M \text{ and } \gamma(t) = p_0\}.$$

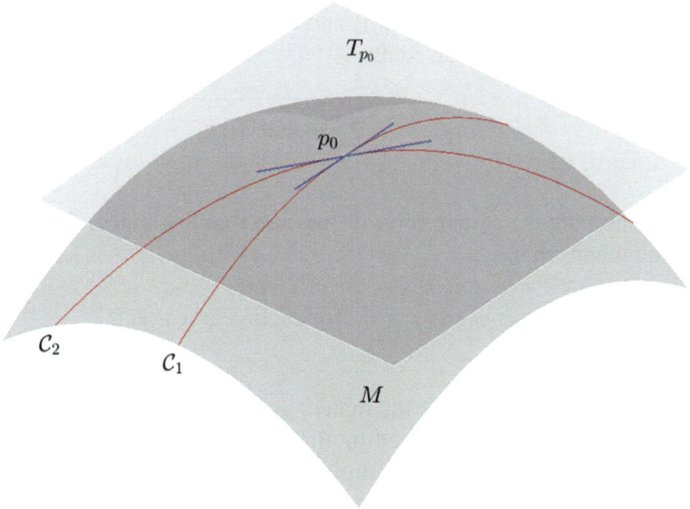

Fig. 3.4 The tangent plane T_{p_0} in a point p_0 is the set of derivatives of curves passing through p_0. It constitutes a vector space and locally approximate the manifold M

The tangent plane to a point p_0, then, is the set of vectors applied to the point p_0 and tangent to M. In this sense, if the integral curve is an observable of the physical world, its derivative is an element of the tangent space, i.e. is an object belonging to the virtual plane.

The tangent plane together with the point p_0 provides a linear local approximation of the manifold. And so, the manifold can be approximated locally by its tangent planes and can even be reconstructed from its tangent planes. As a result, the tangent plane will represent the virtual plane for the displacements on the manifold (Fig. 3.4).

An important property of the tangent plane is the fact that it is a vector space with the same dimension as the manifold. If the manifold has dimension 2, its tangent space will be a plane, if the manifold has dimension 3, its tangent space will be a volume. In general, a manifold of dimension n will have a $n-$dimensional tangent plane at every point.

Since the tangent space at a point p_0 is a vector space, we can introduce a scalar product and a norm on vectors with the aim to measure the length of such vectors and the distance between points.

If we fix a basis of $T_{p_0}(M)$ the length of a vector $\mathbf{v}$ of components $(v_1, v_2, \cdots)$ can be computed as the square root of a second-order polynomial, following the Riemann receipt prescribing that the linear infinitesimal element ds is "expressible by the square root of a second degree differential expression, i.e. the space is a planar quantity in its infinitesimal parts" [12].

Displacements in all directions are allowed starting from a point p_0, with speed expressed by the coefficients $(g_{i,j}(p_0))$. They constitute an invertible $n \times n$ matrix, which we will denote $G(p_0) = (g_{ij}(p_0))$ and defines the norm of the tangent plane $T_{p_0}(M)$. The norm will be expressed as

$$|\mathbf{v}|_{G(p_0)} = \sqrt{\sum_{i,j} g_{i,j}(p_0) v_i v_j}. \tag{3.4}$$

In order to compute the square root, we require that the matrix $G(p_0)$ is positive defined, which means that

$$\sum_{i,j} g_{i,j}(p_0) v_i v_j > 0, \text{ for each vector } v \neq 0. \tag{3.5}$$

For example, in the Euclidean case the matrix $g_{i,j}$ is the identity and displacements are accomplished with the same speed in all directions. If G is diagonal $G(p_0) = (\lambda_1(p_0), \ldots, \lambda_n(p_0))$, this condition simply means that each λ_i is strictly positive. The definition of norm becomes

$$|\mathbf{v}|_{G(p_0)} = \sqrt{\sum_{i}^{n} \lambda_i(p_0) v_i^2}. \tag{3.6}$$

A different weight is given in any direction, meaning that we expect different speeds in different directions. The level set of the norm $|\mathbf{v}|_{G(p_0)} \leq C$ defines the ball on the tangent space. If the norm is Euclidean the ball is a circle (or a sphere). In general it will be an ellipse (or ellipsoid, depending on the dimension). In the diagonal case Eq. (3.6), the highest value of λ is the larger axis. The smallest one is the shorter axis (Fig. 3.5).

3.2.5 *Metrics and Striated Manifolds*

Up to now we have fixed a point p_0 and considered properties of the tangent plane at the fixed point p_0. We can repeat the same construction at every point, obtaining a tangent space $T_p(M)$ at every point $p \in M$. The union of these tangent spaces will be called tangent bundle

$$T(M) = \cup_{p \in M} T_p. \tag{3.7}$$

Each element of the tangent bundle will be a couple (p, v), where p is a point of the manifold and v is a vector at the point p.

If the norm is assigned not only at a fixed point p_0, but it is defined at all points p of the manifold M. , and the function $g_{ij}(p)$ is smooth as a function of the variable p, then g_{ij} defines a metric on M. This means that the norm is allowed to smoothly change from one point to the other, and the corresponding level set of the norm defines smoothly changing ellipses (Fig. 3.6).

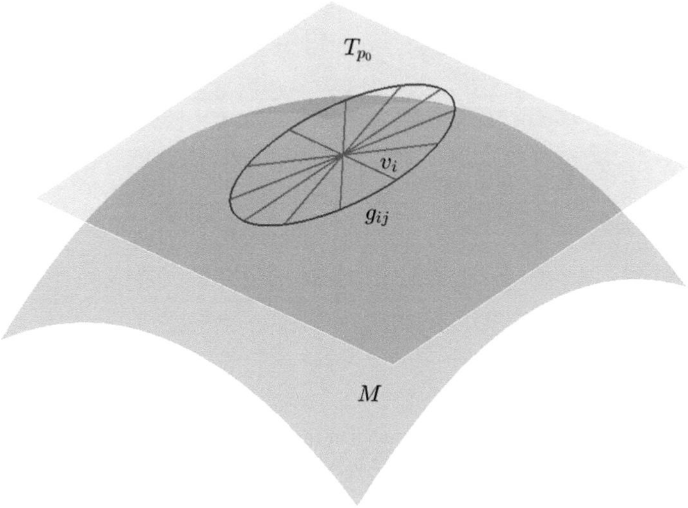

Fig. 3.5 The norm of a vector $\mathbf{v}$ of components $(v_1, v_2, \ldots)$ can be computed as $|\mathbf{v}| = \sqrt{\sum_{i,j} g_{i,j} v_i v_j}$, where $g_{i,j}$ is a matrix weighting the different components. It is namely the metric and it is represented by an ellipse

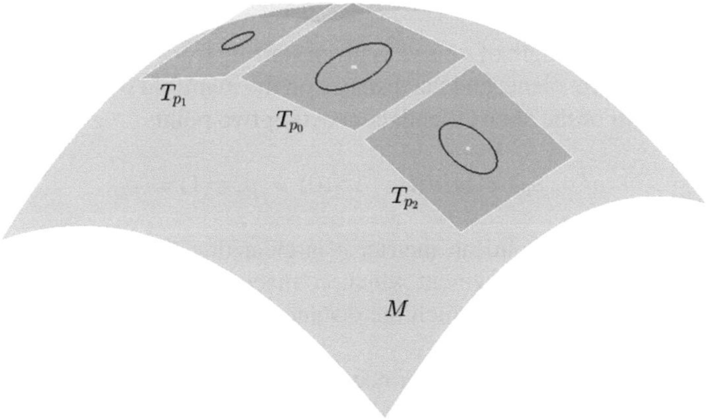

Fig. 3.6 Riemann's spaces are devoid of any kind of homogeneity since the metric can be different point by point

The metric assigns different lengths—always non–zero—to vectors in different directions. If we consider a curve γ, its derivative $\gamma'(t)$ at a point t represents by definition a tangent vector at the point $\gamma(t)$. It belongs to the tangent space, and we need to integrate it to define the length of the curve γ which is an observable object as we already noted. Thus the length is given by

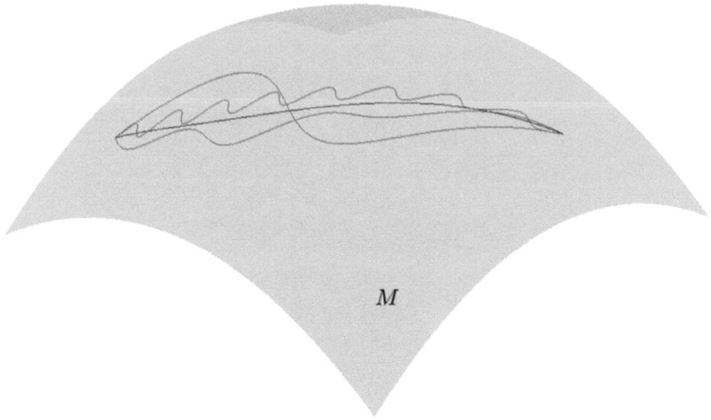

Fig. 3.7 The length of a curve on a Riemannian manifold is given by the shortest path that in general is not a straight line. Note that we are now considering curves, lying on the manifold, not on the tangent planes

$$l(\gamma) = \int_0^1 ||\gamma'(t)||dt.$$

In this manner the length of a curve is given starting from the differential structure defined on the tangent plane, and the distance on the manifold is deduced from there, since is the length of the shortest path connecting two points.

$$d(p, p_0) = min\{l(\gamma) : \gamma(0) = p, \gamma(1) = p_0\}.$$

If we consider the Euclidean metric, it is clear that the distance between two points is the length of the segment which connects them. But this is not the case in a Riemannian geometry in which the distance can be realized by the length of a different curve (Fig. 3.7).

It is now clear that a metric is not necessarily defined on a smooth manifold. If a metric is defined, and consequently it is possible to induce a distance, then the space is called striated. There are beautiful examples of smooth and striated space that Deleuze gives taking inspiration both from music and from cinema.

> Smooth space and striated space - nomad space and sedentary space -the space in which the war machine develops and the space instituted by the State apparatus - are not of the same nature ([3] (1980): 473).

> The striated is that which intertwines fixed and variable elements, produces an order and succession of distinct forms, and organizes horizontal (melodic) lines and vertical (harmonic) planes. The smooth is the continuous variation, continuous development of form; it is the fusion of harmony and melody in favour of the production of properly rhythmic values, the pure act of the drawing of a diagonal across the vertical and the horizontal (ibid., 477).

> In a smooth space-time, one occupies without counting, whereas in a striated space-time one counts in order to occupy (ibid., 478)

It's important to note that there can be heterogeneity both in smooth and in striated spaces. In particular, the heterogeneity of smooth spaces is due to the infinite variety of charts with which the manifold can be covered while the heterogeneity of striated spaces is due to the possibility of changing the norm point by point.

Let's note that independently of the smooth or striated character of a manifold, a manifold fulfils the three conditions of a multiplicity, which we've previously considered and now recall: (1) virtuality—since tangent spaces are collections of vectors, that is derivatives, (2) juxtaposition—since the manifold is constructed by a set of charts (3) differentiability—since charts are glued under the condition of continuity of the first derivative.

Let's now extend this understanding of multiplicity to the concept of assemblage, which latter accounts for the composition of any multiplicity.

3.3 The Plane of Composition

Jean Petitot, in his essay "Morphogenesis of meaning" [11], has offered a remarkable analysis of the manner in which Thom's and Deleuze's mutually independent works demonstrate close resemblances: each attempts to realize structural objectivity in terms of a geometry of positions and thus to fulfil the Lautman's programme for singularities. Each of these aspects is summarized by Petitot in a very short and intense sentence:

> Every structure has the following two aspects: a system of differential relations according to which the symbolic elements are determined reciprocally, a system of singularities corresponding to these relationships and plotting the space of the structure.[1]

While Petitot outlined similarities between the programmes of Thom and Deleuze, we would like to stress here an important difference between the two. Thom's structural morphodynamics, as we've considered above, is essentially a theory of control of singularities, in such a way that a set of state variables can control the system and drive it towards desired stable states. As such, the internal state space is given a priori. In Deleuzian dynamics, on the other hand, singularities are primary with respect to parameter space and they are generated in such a manner as to constitute a new constellation of points:

[1] Toute structure présente les deux aspects suivants : un système de rapports différentiels d'après lesquels les éléments symboliques se déterminent réciproquement, un système de singularités correspondant à ces rapports et traçant l'espace de la structure.

> To think is to emit singularities. If it were a definition of thought, we would better understand Mallarmé's "A dice ...", we would better understand Nietzsche's call to the game of dice. To think is to roll a dice. What are singularities? These are dots on the faces of the dice coming out." [2][2]

Then it's not anymore a matter of *control* of a given set of singularities, as it was in catastrophe theory, but rather to *create* constellations of singularities: "To think is to throw a dice, that means, once again, well yes, that chance itself is a balance of power".[3]

If in the logic of control the structure contains rules of transformation from a state to another of a system, in the dynamic of imagination the space of possibility has to change constantly until all the rules have constantly changed.

To allow a mutation of the space of possibilities, it is necessary a transformation of the virtual that generates it, which goes well beyond the parametric variation. As long as we are in the regime of parametric variation we remain within the same space of possibilities, where the mathematical framework proposed by R.Thom is well suitable to take into account for example phase transitions. But to allow the transformation of the phase space and introduce a new one it is necessary to take into account a recomposition of the virtual that is not common to mathematical physics and either in structural morphodynamics.

The virtual plane has to become a true plane of composition where differential fields are recombined. The imaginative process is played on this composition of force fields that are recombined hoping that they work, that is, that they integrate to unleash a new form or a new idea. To achieve integrability it is necessary to add fields that complete the existing ones. Deleuze defines these fields as "adjuncts" because they constitute the complementary conditions to make the problem solvable. Very concisely there are two concurrent processes which intervene in the logic of imagination: the specification of adjunct fields on the virtual plane and the condensation of singularities in its actualization.

> On the one hand, in the progressive determination of the conditions, we must in effect discover the adjunctions which complete the initial field of the problem as such—in other words, the varieties of the multiplicity in all its dimensions, the fragments of ideal future or past events which, by the same token, render the problem solvable; and we must establish the modality in which these enclose or are connected with the initial field. On the other hand, we must condense all the singularities, precipitate all the circumstances, points of fusion, congelation or condensation in a sublime occasion, Kairos, which makes the solution explode like something abrupt, brutal and revolutionary. ([6] (1968)).

Once again we outline that differential calculus is not considered here as an abstract computational tool but it is a genetic device embedded in natural dynamics, embodied in the becoming of corps and extended to social practices:

---

[2] "Penser c'est émettre des singularités. Si c'était une définition de la pensée, on comprendrait mieux Un coup de dés... de Mallarmé, on comprendrait mieux l'appel de Nietzsche au jeu de dés. Penser c'est émettre un coup de dés. Les singularités, c'est quoi ? C'est des points sur les faces du dé qui sort. Les faces sortantes du dé." [2]

[3] "Penser c'est émettre un coup de dés, ça veut dire, encore une fois, ben oui, que le hasard lui-même est un rapport de forces". [2]

> It is in this sense that Lenin had Ideas. (There is an objectivity on the part of adjunction and condensation, and an objectivity of conditions, which implies that Ideas no more than Problems do not exist only in our heads but occur here and there in the production of an actual historical world.) Furthermore we must not see mathematical metaphors in all these expressions such as 'singular and distinctive points', 'adjunct fields', and 'condensation of singularities', nor physical metaphors in 'points of fusion or congelation ...', nor lyrical or mystical metaphors in 'love and anger'. These are categories of the dialectical Idea, the extensions of the differential calculus (mathesis universalis but also universal physics, universal psychology and universal sociology) corresponding to the Idea in all its domains of multiplicity. They are what is amorous or revolutionary in every Idea, that by virtue of which Ideas are always unequal glimmers of love and wrath which have nothing in common with any natural light. ([6] (1968))

When the composition of differential fields is actualized, different possible scenarios open up. If the flow is trapped in stable basins we see that a new structure can emerge under the condition that at least two basins are present and a device allowing the passage from one basin to another is available. In this case, the concept of structural dynamics is extended to that of structural event, meaning the passage from one structure to another one. This interpretation has been recently taken up: "But if an event cannot be reduced to a change of facts then how can it be determined? Well, precisely as a change in structure. The concept of structure has never had the aim or the effect of making the very experience of change doubtful, but on the contrary of distinguishing between two types of change: changes that do not affect the structure, and that can be called facts, and changes that affect structure, and that will be called events" (P. Maniglier La structure de l'événement, 2020, personal communication).[4]

Anyway, the structural event is unable to get rid of oppositional and even binary relationships, which structures instantiate between categories in terms of bi-univocal relationships between positions. The structural event is the passage from one structure to another one by preserving the deep functioning of the structure as a device for categorization. René Thom has shown clearly that the emergence of structures requires a very particular kind of singularities defining stable basins of attraction. Indeed, the Thomian theory of structural stability in its entirety aims to pose the conditions for the stabilization of fluxes. Mathematically, this means that attractor basins are stable and oppositive relations that are implemented by means of parameters changing in an external space. These are the condition of existence for a structure.

But of course there are other possible actualizations of the composition of adjoint fields including general Poincaré singularities that differ from gradient potential. Likewise, there are non-standard attractors like strange attractors, as well as solutions coming from harmonic analysis such as those at work in quantum mechanics,

[4] "Mais si un événement n'est pas réductible à un changement de faits, alors comment le déterminer? Eh bien précisément comme un changement de structure. Le concept de structure n'a jamais eu pour but ni pour effet de rendre douteuse l'expérience même de changement, mais bien au contraire de distinguer entre deux types de changements: des changements qui n'affectent pas la structure, et qu'on peut appeler des faits, et des changements qui affectent la structure, et qu'on appellera alors des événements" (P. Maniglier La structure de l'événement, 2020, personal communication).

which were considered in detail by Gilbert Simondon. In this latter case, the actualized flow and its virtual counterpart become progressively more autonomous from structures. The emerging dynamics do not necessarily give rise to stabilized singularities, with their oppositional relations, and it remains in the form of an unconstrained flow, eventually giving rise to plateaus, meaning a plurality of modes of vibrations resulting from the harmonic analysis of the flow. We use here the word "plateau" as Deleuze and Guattari do in "A thousand plateaus", that is in the sense with which Gregory Bateson indicates a self-vibrating region of intensities that do not cut the space but rather coexisting therein. The genesis of the imaginative event finds its focus in the composition of virtual fields and particularly in the constitution of "agencements", that is to say, heterogeneous differential assemblages which eventually can be integrated.

In a post-structural logic of imagination, the composition of adjunct fields that form assemblages thus completely replaces the structural device. In the assemblage, relationships multiply to become heterogeneous in the form of connections, conjunctions, alliances. An assemblage is therefore not a changing structure, nor is it necessarily the virtuality of a structure. Simply put, it is not a structure and its actualization does not necessarily produce a structure. Its internal and external relations have nothing to do with the oppositional relations of a structure. Both structural and post-structural dynamics play on the axis of an external time: the time of changing parameters for Thom, the time of the composition of assemblages, the Aion, for Deleuze. *Notice that the practice of plastic composition is not limited to the addition of differential fields but may imply a multiplicity of actions including substitution, changing, cutting, removal, and any other pragmatic action of transformation.*

An assemblage is thus an a-structural genetic element for any dynamic individuation, the composition and actualization of which give rise to differential heterogenesis. We will consider in what follows the principal mathematical arguments used by Deleuze and Guattari to define the notion of assemblage as a composition of multiplicities.

3.4 Assemblages

A thousand plateaus opens with an introduction entitled Rhizome, in which Deleuze and Guattari describe the history of western philosophy as being comprised of disjunctions, or ... or ... or .... They propose instead a philosophy of conjunctions, and ... and ... and ... .

The conjunctions they envision can only occur between multiplicities, since everywhere there are only multiplicities or multiplicities of multiplicities. From here, the assemblage comes to be conceived of as the conjunction of multiplicities: "packs in masses and masses in packs" ([3] (1980)). All empirical reality is understood to be the product of an assemblage, reality itself existing only by way of the connection of multiplicities: "An assemblage is precisely this increase in the dimensions of a multiplicity that necessarily changes in nature as it expands its connections" ([3] (1980)).

The conjunctive logic of and ... and ... and ... is put in place by the immanent dynamic of the assemblage - that is, by the establishing of heterogeneous connections between multiplicities: "What is an assemblage? It is a multiplicity which has many heterogeneous terms, and which establishes links, relations between them, through ages, sexes, kingdoms—different natures. Also, the only unit of the arrangement is co-operation: it is a symbiosis, a 'sympathy'" ([5] (1977): 84).[5]

Assemblages are in constant variation; they are themselves constantly subject to transformation. In this sense, they must be recognized as eventual configurations and real inventions. The actualization of an assemblage gives rise to the emergence of a flow, with its particular singularities—but this isn't all; it also gives rise to something else. In considering the actual elements of an assemblage, Deleuze and Guattari specify that an assemblage is made of plateaus, which, we reiterate once more, is to be understood in the sense in which Gregory Bateson uses the word: a "plateau designates something very special: a continuous, self-vibrating region of intensities whose development avoids any orientation towards a culmination point or external end" ([3] (1980)). In the Balinese culture studied by Bateson, particularly in music, plateaus of intensities build up and fall away, shift and transform without any linear structure. These self-vibrating regions attest to the fulfilment of the condition of consistency of the assemblage and are co-extensive in space, in the sense that they overlap themselves in the same region, instead of partitioning the space as structural categories or basins of attraction à la Poincaré.

The actualization of an assemblage is made by harmonic modes of vibrations, which are capable of mapping the space in a continuous and supply fashion rather than cutting it into categorial domains comprised of binary oppositions. A tensive space is thus built by the set of continuous self-vibrating regions. We will see in Chap. 4 that this definition of plateaus as self-vibrating regions will open onto a mathematical interpretation in terms of the eigenvectors of an affinity matrix, which is at the base of the constitution of neurogeometrical morphologies in the study of brain organization as well as of tensive spaces in primary semiosis.

In this formidable Deleuzo-Guattarian invention, we cannot help but notice slightly different sensibilities at work: one more geometrical, the other more dynamical as concerns the definition of the assemblage (we avoid of course assigning these respectively to Deleuze and Guattari). In other words, the term *assemblage* in A thousand plateaus is declined both from a geometric point of view, in terms of *rhizome*, and from a dynamic perspective, in terms of *machine* or *machinic assemblage*, that

---

[5] Qu'est-ce qu'un agencement ? C'est une multiplicité qui comporte beaucoup de termes hétérogènes, et qui établit des liaisons, des relations entre eux, à travers des âges, des sexes, des règnes—des natures différentes. Aussi la seule unité de l'agencement est de cofonctionnement : c'est une symbiose, une "sympathie".

is, the heterogeneous becoming of forms from a primordial chaotic soup. Even if this distinction is not prescriptive, we will maintain it as it uniquely allows us to better articulate the geometric and dynamic aspects of the assemblage.

3.5 Rhizomes

Assemblages and rhizomes are nearly synonymous; they refer to the same object, though each entails a different emphasis or shading. Whereas the former refers primarily to the process through which multiplicities are conjoined, the latter references the topology of spaces continuously created by way of this same process of assembly. Accordingly, on the basis of the definitions we've just provided, rhizomes and rhizomatic multiplicities can be understood to deal with the topological aspect of the assemblage.

Deleuze and Guattari outline the concept of the rhizome in the introduction of A Thousand Plateau, listing the principles of its construction

> 1 and 2. Principles of connection and heterogeneity: any point of a rhizome can be connected to any other, and must be...; 3. Principle of multiplicity: it is only when the multiple is effectively treated as a substantive, "multiplicity", that it ceases to have any relation to the One... 4. Principle of a-signifying rupture: against the oversignifying breaks separating structures or cutting across a single structure. A rhizome may be broken, but it will start up again on one of its old lines, or on new lines... 5 and 6. Principle of cartography and decalcomania: a rhizome is not amenable to any structural or generative model; it is a map and not a tracing.

The rhizome is thus introduced as an a-centred system mainly in opposition to the hierarchical structure of the tree or of the root: "In contrast to centred (even polycentric) systems with hierarchical modes of communication and pre-established paths, the rhizome is an a-centred, nonhierarchical, nonsignifying system without a General and without an organizing memory or central automaton, defined solely by a circulation of states" ([3] (1980)).

However, a rhizome can nonetheless contain such structures, taking the form of a complexification thereof: "There exist tree or root structures in rhizomes; conversely, a tree branch or root division may begin to burgeon into a rhizome. The coordinates are determined not by theoretical analyses implying universals but by a pragmatics composing multiplicities or aggregates of intensities. A new rhizome may form in the heart of a tree, the hollow of a root, the crook of a branch. Or else it is a microscopic element of the root-tree, a radicle, that gets rhizome production going".

Structures like trees or roots are characterized by rigid segmentarities, meaning dichotomic relations, whereas rhizomes allow a more supple segmentarity:

> We may summarize the principal differences between rigid segmentarity and supple segmentarity. In the rigid mode, binary segmentarity stands on its own and is governed by great machines of direct binarization, whereas in the other mode, binarities result from "multiplicities of n dimensions." And again rigid "segmentarity feeds into a machine of overcoding that constitutes more geometrico homogeneous space and extracts segments that are determinate as to their substance, form, and relations. It is worth noting that this rigid segmentarity is always expressed by the Tree. The Tree is the knot of arborescence or principle of dichotomy; it is the axis of rotation guaranteeing concentricity; it is the structure or network gridding the possible. This opposition between arborified and rhizomatic segmentarity is not just meant to indicate two states of a single process, but also to isolate two different processes.

There are many important differences between a rhizome and a structure.

While the structure is made of points and positions, the rhizome is made of relational lines. While relations between the points of a structure are of the oppositional-binary type, the assemblage has a heterogeneous relational set:

> Unlike a structure, which is defined by a set of points and positions, with binary relations between the points and biunivocal relationships between the positions, the rhizome is made only of lines: lines of segmentarity and stratification ([3] (1980): 22).

A rhizome, then, is not to be understood as a new structure, but nor is it an ever changing structure. Its internal and external relations have nothing to do with the oppositional relations of a structure:

> It is certain that they have nothing to do with a structure, which is never occupied by anything more than points and positions, by arborescences, and which always forms a closed system, precisely in order to prevent escape.

Moreover, while the structure is characterized by points of stability (such as sink singularity) or of stabilization (such as catastrophes à la Thom), the rhizome is constituted by flows that eventually condense into modes of vibration:

> a flow of children; a flow of walking with pauses, straggling, and forward rushes; the semiotic flow of the confessions of all the children who go up to the old monk at the head of the procession to make their declarations; a flow of desire and sexuality. ... What is important is not whether the flows are "One or multiple"–we're past that point: there is a collective assemblage of enunciation, a machinic assemblage of desire, one inside the other and both plugged into an immense outside that is a multiplicity in any case. ([3] (1980)) pag 23.

The rhizome is continuously linked with an immense series of externalities, which serve to increase the size of the rhizome. If the term structure names the systemic organization par excellence, the assemblage is the differential of a xeno-systemics in which exteriority continually changes its non-arborescent organisation in such a manner as to undergo a metamorphosis:

> When a multiplicity of this kind changes dimension, it necessarily changes in nature as well, undergoes a metamorphosis. ([3] (1980): 23)

"The rhizome operates by variation, expansion, conquest, capture", but also it can split and divide itself. It is the operator of both the conjunctive linkage of "and", on the one hand, and of schismogenesis, on the other, that is, the analog of a cut on the virtual plane.

Many of Deleuze's readers have interpreted the rhizome as a network, often glimpsing therein the announcement of telematic networks. However, even if it's the case that the topology of the network escapes that of the tree, the structure, and the root, it nonetheless remains insufficient as an explanation of material assemblages, which are characterized by indeterminacy and inaccuracy. The heterogenous relations of the assemblage have to be considered as "directions in motion", that is directions of vector fields that are dense and that contain all the uncertainties of material becoming.

Non-commutative vector fields induce an uncertainty principle that generalizes the Heisenberg uncertainty principle to observables other than position and momentum. In this sense they have little to do with a network and even less with a digital network.

A principle of material conjunction or alliance, rather than a principle of logical connection, would seem to be at the ground of rhizomatic propagation.[6] On this basis, we would suggest that the most appropriate definition of a Rhizome would seem to be the geometric counterpart of an assemblage, that is a manifold. We will return to this concept in the following chapters.

3.6 Machines

When outlining the dynamic dimension of the assemblage, Deleuze and Guattari frequently employ the term "machinic assemblage", or even, more directly, machine:

> We think the material or machinic aspect of an assemblage relates not to the production of goods but rather to a precise state of intermingling of bodies in a society, including all the attractions and repulsions, sympathies and antipathies, alterations, amalgamations, penetrations, and expansions that affect bodies of all kinds in their relations to one another ([3] (1980): 90).

The term machinic assemblage is not used in the sense of an algorithm or an automatism, but as a synonym for a heterogeneous process, with heterogeneity here pertaining to intensive relations that exceed a pure connective function, becoming instead both multiple and conjunctive.[7]

The machine can thus be understood as the extension of the geometric concept of rhizome to the field of dynamics. Far from being limited to the status of a technological device, a machine is the process capable of deploying something new, or of producing a singular individuation. The machinic process is heterogenetic in the sense that it interprets the Simondonian process of individuation in a completely heterogeneous fashion, as Guattari explains in his article about machinic heterogenesis:

> A creationist conception of the machine - whether scientific, theoretical, aesthetic, or informational - reconsiders the philosophical foundations of ontology. It is then no longer a

---

[6] On the concepts of conjunction and connection see also the PhD thesis of Franco Berardi "AND, phenomenology of the end" [1].

[7] On the notion of machine and mechanism see also the volume of Ubaldo Fadini, "Soggetto e fantasia. Per un'antropologia macchinica" [7].

> universal enunciator but a multiplicity of partial subjectivities which never cease to assert themselves in their heterogenesis. These operators have an ethico-political scope, because they put the emphasis back on possible praxis, whether in the domain of affects, percepts or concepts. [10][8]

And again:

> The structure is interactional, involves feedback loops, brings into play a concept of totalization that she masters on her own. It is inhabited by inputs and outputs which are intended to make it function according to a principle of eternal return. She is haunted by a desire for eternity. The machine, on the contrary, is wrought by a desire for abolition. Its emergence is doubled by the breakdown, the catastrophe, the death which threaten it. It has an additional dimension: that of an otherness that it develops in different forms. And this otherness separates it from the structure, centered on a principle of ho-meo-morphy. The difference brought about by machinic autopoiesis is based on imbalance, the prospecting of virtual universes far from equilibrium. And it is not only a question of a rupture of formal balance, but of a radical ontological reconversion [10].[9]

In this sense, there cannot be a homogeneous Being. Guattari lashes out against the Heideggerian homogenetic ontology: Heidegger "sinks into a Being who is for himself a chaosmic vertigo. So, obviously, if the Being corresponds to this chaosmic vertigo, everything collapses in a loss of radical sense and in a catastrophic pessimism". [10] [9]. And even more importantly: what looms behind beings is therefore not a homogenesis. Rather, guaranteeing the multiplicity of beings and things is the primordial origin of heterogeneous ontological dimensions.

The idea of a heterogenetic machine was born in the psychoanalytic field with the intention of defining the unconscious without the use of predefined categories. It has since been developed, however, in the direction of an ontological transversality. If the machinic unconscious would remain at the centre of Guattarian research, the heterogeneous meta-model derived from this conceptual origin would soon become the dynamic of becoming for other domains as well. We can see this in "What is philosophy" ([4] (1991)), in which the machinic unconscious is extended towards a cog-

---

[8] "Une conception créationniste de la machine - qu'elle soit scientifique, théorique, esthétique ou informationnelle - reconsidère les fondements philosophiques de l'ontologie. Il n'est plus alors d'énonciateur universel mais une multiplicité de subjectivités partielles qui ne cessent de s'affirmer dans leur hétérogenèse. Ces opérateurs ont une portée éthico-politique, car ils remettent l'accent sur des praxis possibles, que ce soit dans le domaine des affects, des percepts ou des concepts."

[9] "La structure est interactionnelle, implique des boucles de rétroactions, met en jeu un concept de totalisation qu'elle maîtrise à partir d'elle-même. Elle est habitée par des inputs et outputs qui ont vocation à la faire fonctionner selon un principe d'éternel retour. Elle est hantée par un désir d'éternité. La machine, au contraire, est travaillée par un désir d'abolition. Son émergence est doublée par la panne, la catastrophe, la mort qui la menacent. Elle possède une dimension supplémentaire : celle d'une altérité qu'elle développe sous différentes formes. Et cette altérité l'écarte de la structure, axée sur un principe d'ho-méo-morphie. La différence apportée par l'autopoèse machinique est fondée sur le déséquilibre, la prospection d'univers virtuels loin de l'équilibre. Et il ne s'agit pas seulement d'une rupture d'équilibre formel, mais d'une radicale reconversion ontologique."

[10] est le tenant d'une ontologie homogénétique. Il sombre dans un Être qui est pour lui-même un vertige chaosmique. Alors, évidemment, si l'Être correspond à ce vertige chaosmique, tout s'effondre dans une perte de sens radical et dans un pessimisme catastrophique.

nitive unconscious that involves heterogeneous brain dynamics and that opens onto the question of "thought as heterogenesis". The novelty of the Guattarian paradigm, however, is that it opens more widely still onto an extended and multiple ecologies containing all kinds of environmental, social, and mental virtuality.

Here, the heterogenetic machine becomes the model for the pre-individual field involved in an extended morphogenetic becoming. The fundamental reason for this precipitous expansion of the concept is that the great machine, which can never be normalized, is nothing other than nature itself. Nature captured in the aspect that Spinoza would have called naturans, and which includes within itself that to which it is typically opposed, namely, human technics.

Nature, in other words, is the machine: as natura naturans it is life as the process of living.

We can therefore recognize that it is characteristic of machinic heterogenesis to escape both from structure and from chaos, since neither the one nor the other is capable of constituting an event. Both are too predictable, structure due to an excess of rigidity, chaos for a lack of consistency. It is thus in opposition to structures that the heterogenetic machine is understood to perpetually transform the very rules of its own operation, exceeding thereby any oppositional duality; in this aspect, it operates as a perpetual morphogenesis that destroys rules at the same time that it invents new possibility spaces. For this reason, the heterogenetic machine exemplifies the dynamic of Guattarian individuation, with its double character expressed by the concept of chaosmosis ([8] (1992)). Chaosmosis, originally introduced in the work of James Joyce, is a mixture of Chaos and Cosmos wherein we attend to the emergence of Cosmos from the chaotic action of multiplicities that destroy old rules and recombine new ones by way of heterogeneous relations that overcome any categorial duality of structures. In this sense, any anarchic and a-categorial virtuality cannot possibly cease to create new forms; its very essence is to do precisely this.

With all of this taken into consideration, a question remains as to the possibility of integrating these two perspectives on the assemblage—its geometric aspect, which we've explored in the concept of rhizome, and its dynamic aspect, which we've considered by way of the concept of machine. In what follows, we will take up precisely this question, as well as, more generally, the question: how is it that the intuition of heterogeneous machines can reinterpret the process of differential becoming from a morphodynamical point of view?

In other words, we will interrogate the possibility of conceiving heterogenesis in light of contemporary mathematics.

3.7 Towards New Geometries and Dynamics

It is important to note that Riemannian differential geometry is able to capture many aspects of the rhizome. For example: As an assemblage is the conjunction of different multiplicities, so a manifold is defined by a conjunction of local patches.

Since a multiplicity is precisely a manifold, as we have seen at the beginning of the chapter, it follows that an assemblage is a manifold, considered as an union of heterogeneous patches. Furthermore the heterogeneous character of the assemblage is partly fulfilled by the Riemannian manifold, considered as a juxtaposition of heterogenous patches, where heterogeneity is given either by the variety of ways to glue patches in smooth spaces or by the variety of metric tensors that can be chosen in striated spaces.

In addition, the Riemannian metric is able to capture one of the main properties of the rhizome: its complete connectivity: "any point of a rhizome can be connected to any other" ([3] (1980): 7). In fact a differential manifold in the sense of definition (3.2) and (3.3) is connected if it satisfies the following additional condition: For every point x and y of the manifold M there exists a curve lying on the manifold connecting the two points.

If the space is smooth, this property is fulfilled with topological arguments and if the space is striated this property at the basis of the definition of distance, meaning that the distance between the two points is finite.

However, it is important to note that a Riemannian manifold as a gluing of patches is not sufficient to define an assemblage. In fact, as stated by Deleuze and Guattari, the definition of assemblage implies a change of dimension and of nature of the space: Indeed "an assemblage is precisely this increase in the dimensions of a multiplicity that necessarily changes in nature as it expands its connections" ([3] (1980)).

This increasing of dimension and changing in nature is unquestionably one of the most important features of the concept of assemblage.

In contrast, gluing together patches of a manifold does not introduce change in dimensions or in nature. The dimension of a Riemannian manifold and its tangent space is uniform at every point, and this is its constitutive property, in contrast with the rhizomatic property in which space can change in both dimension and nature in a molecular way: "When a multiplicity of this kind changes dimension, it necessarily changes in nature as well, undergoes a metamorphosis" ([3] (1980): 23).

The change of dimension and nature cannot take place otherwise than through a change of the tangent bundle, that is the space of possibility of the system. An assemblage, then, instantiates a double morphogenesis: a becoming of forms, as transformations inside a space, and a becoming of the associated possibility spaces. In an assemblage, the possibility space becomes molecular—that is, local—and changes on a point to point basis in space as well as in time. The concept of Riemannian manifold thus has to be extended, in order to take into account this new molecularity, which allows for a morphogenesis of tangent spaces.

New geometries are also needed to take into account the complexity of the concept of differential put forward by Deleuze. Consider the example offered by the process of integration: the constellations of the virtual are transformed into forms when they find a common frequency, a morphogenetic vibration that actualizes them in doing so, however, they bring with them pieces that do not integrate but rather resist, preserving a dissonant tension into the field, like fields of tension that resist. And here Deleuze, overcoming Leibniz's concept of differential, as we considered above in, indicates that, though there are indeed convergent series that are actualized here and now, there are also divergent ones that will never be integrated, at least not in this world. Such an understanding, it's worth noting, may well be taken from the Simondonian conception of individuation, as a process of always partial integration in which the pre-individual resists taking shape, often preferring the intensity of power over the extension of form. And so, if, in the Riemannian geometries of assemblages developed by Deleuze and Guattari in A thousand Plateaus, everything is connected with everything, new geometries are needed, geometries that would provide a an understanding of a virtual that can be cut, that can be fragmented, and that retains a degree of resistance to the great binge of the actual. In other words, a new dialectics of forces and forms must occur, and specifically, by way of geometries less regular than the Riemannian. It's this that would open onto new dimensions of composition of the virtual.

If phase spaces have to be strongly heterogenized allowing a change of dimension and nature on a point-by-point basis, a similar fate concerns the dynamics that inhabit these spaces of possibilities.

The "ontological heterogenesis" introduced by Guattari in Chaosmosis defines precisely this dynamic heterogeneity, that is a molecular dynamics in continuous recombination and capable to give rise to a fluid matter engaged in singular processes of individuation. For Guattari, individuation is a "machinic subjectivation" that directly brings into play the heterogeneous virtual of dynamic constraints. Such a perspective reconsiders and extends the concept of molecularity that was already introduced in A thousand plateaus: "If the effects realize something this is because the relations between forces, or power relations, are merely virtual, potential, unstable, vanishing and molecular, and define only possibilities of interaction, so long as they do not enter into a macroscopic whole capable of giving form to their fluid matter and their diffuse function" ([3] (1980): 37).

Finally, we should consider processes instantiated by local geometry and dynamics changing point by point on a manifold, in such a way to define a *heterogeneous point-to-point dynamic inhabiting local possibility spaces*, far from uniform dynamics in global phase spaces of mathematical physics and structural morphodynamics.

References

1. Berardi, F.: And: Phenomenology of the End. MIT Press, London (2015)
2. Deleuze, G. cours du 17/12/1985. Sur Foucault Les formations historiques', Transcription: Annabelle Dufourcq (avec l'aide du College of Liberal Arts, Purdue University)
3. Deleuze, G., Guattari, F.: A thousand plateaus. In: Capitalism and Schizophrenia, vol. 2. University of Minnesota Press, Minneapolis, 1987 (1980)
4. Deleuze, G., Guattari, F.: What is Philosophy? Verso, London, 1994 (1991)
5. Deleuze, G.: A qua reconnait-on le structuralisme? In: Châtelet, F. (ed.) Histoire de la philosophie VIII, Verviers Marabout (1973)
6. Deleuze, G.: Difference and Repetition. Columbia University Press, New York, 1968 (1994)
7. Fadini, U.: Soggetto e fantasia. Per un'antropologia macchinica, Clinamen, Firenze (2020)
8. Guattari, F.: Chaosmosis, An Ethico-Aesthetic Paradigm. Trans. Paul Bains and Julian Pefanis. Indiana University Press, Bloomington & Indianapolis, 1995 (1992)

9. Guattari, F.: Vertige de l'Immanence, Entretien, Chimeres, n.38 (1992)
10. Guattari, F.: L'hétérogenèse machinique, Chimères. Revue des schizoanalyses, Annèe **11**, 78–97 (1991)
11. Petitot, J.: In: Brandt, P.-A., Wildgen, W. (eds.) Morphogenesis of Meaning. Peter Lang, Bern, 2004 (1985)
12. Riemann, B.: Über die Hypothesen, welche der Geometrie zu Grunde liegen, Habilitationsschrift, 1867 (1854)

Chapter 4
Differential Heterogenesis

4.1 Discussing Homogenesis

4.1.1 Geometric and Dynamic Heterogeneity

As we have seen in the previous chapter, manifolds are heterogeneous multiplicities capable of being joined together and recombining in new assemblages. However, Riemannian manifolds are in some way limited with respect to the kind of heterogeneity they are capable of instantiating, since they are lacking certain crucial features that would enable a properly heterogeneous differential becoming; more specifically, the major limitation in question consists in the fact that the manifold has the same dimension in every point of the space. In this sense, the differential constraints imposed by the metric, although different, result homogeneous in its dimension as the tangent planes have the same dimension point by point. If we consider the classical problem of integration of a curve starting from its tangents, the differential constraints are homogeneous, or more precisely, equiregular; however, nothing prevents from leaving more freedom to differential operators and considering heterogeneous constraints. To overcome these limitations, we freely interpret the heterogeneity from at least two different perspectives. At a first level, we find heterogeneity in the constitutive variation of differential constraints, which can induce a variety of dynamical behaviours that changes from point to point in such a manner that, once actualized, it will give rise to fluxes. A second level of heterogeneity is also present, since each differential constraint has its own structure of tangent planes that constitutes the possibility space, on which fluxes are allowed to flow.

The differential problem is therefore posed in terms of a composition of heterogeneous differential constraints that form assemblages. Heterogeneous assemblages, meanwhile, have to be built not on the basis of a logic of compatibility or compliance, but by the possibility of creating new spaces and new dynamics that are not given a priori. In this manner, phase spaces as well as dynamics are invented by the intrinsic construction of the singular composition.

© Springer Nature Switzerland AG 2022

A. Sarti et al., *Differential Heterogenesis*, Lecture Notes in Morphogenesis,
https://doi.org/10.1007/978-3-030-97797-9_4

The mathematical problem we will now consider pertains to how it is that this heterogeneous composition is in fact feasible. In the next sections, we will try to clarify how it is that the conjunction of heterogeneous differentials is able to give rise to an assemblage. To do so, we will start from an extension of the" concept of multiplicity, moving from the Riemannian to the sub-Riemannian setting. We will see that in the new setting of sub-Riemannian geometry, thanks to the huge contribution of Hörmander [15] and Rothschild & Stein [26] a change of dimension and nature of a manifold is allowed, eventually varying point by point. A similar process of heterogenization of differential constraints can be applied to the dynamical counterpart of assemblages. This is the premise that will allow to construct assemblages by adding, changing, subtracting, cutting, eliminating differential fields.

4.1.2 Beyond Mathematical Physics

We've begun to take quite a distance from the usual differential calculus of mathematical physics, in which the distributions of operators remain spatially and temporally homogeneous. In heterogenesis, there is a spatially and temporally varying definition of differential constraints. Mathematical physics is a form of symmetrization of heterogenesis in the sense that any heterogeneous set is reduced to a unique operator that holds in every spatio-temporal point. Heterogenesis, then, can be regarded as a hyperphysics that takes place as a variety of dynamics flowing on a multiplicity of tangent bundles which change molecularly from point to point.

This character of "homogenization" of mathematical physics is at the basis of its fundamental a priori: the presupposition that spaces are given in advance with respect to differential constraints. Such an assumption is completely reversed in the composition of heterogenetic assemblages, in which operators are primary and serve to define both the dimensions and the qualities of the space. A new differential field that is composed along with an assemblage thus redefines completely the spaces of the entire assemblage.

In mathematical physics, operatorial homogeneity and the fixity of the differential constraints determine together the universality of laws and the nomological character of differential models for an immanent morphogenesis that is created on a case by case basis by the assembly of singular concatenations of sub-Riemannian manifolds.

Notice that if the assemblage of operators is considered in turn a new differential operator, heterogenesis can be viewed as a morphogenesis of the assemblage operator itself. The heterogenetic becoming is then considered a concurrent morphogenesis of operators, of its spaces and of forms in spaces. Such a conception is unprecedented in both physical and structural dynamics.

4.1.3 Khronos and Aion

To allow for the construction of assemblages, two temporal scales or axes will be introduced. The first one is the axis of the actualization of differential constraints. It is the axis of Khronos, which is common to mathematical physics. The second is the axis that Deleuze calls Aion, along which the recombination of differential constraints into new assemblages takes place.[1] On this axis, we have a true plasticity of the virtual, which is to say, we find the possibility to recombining genetic elements so as to create singular dynamics.

Any specific composition has to be thought of as an explorative action, closer to a Dada performance than to a finalized process. The act of composition of forces is not subjected to any mathematization, nor to any other pre-established rule; behind the act of composition, there is only the concreteness of the gesture.

The composition of a singular assemblage, then, is of necessity an invention; it is the creation of new dynamics, instant by instant. The inventive character of the assemblage is due to the fact that the space created by the assemblage is much more than the union of identitary spaces of single operators.

4.1.4 Mathematical Constructivism and Historical Contingency

What is the purpose of reconsidering differential heterogenesis from a morphodynamical point of view today?

The first motivation relies on the fact that the very idea of the becoming of forms has an operational nature. Becoming assumes from the beginning a problematic dimension, in the strict mathematical sense of posing and (re)solving a problem. The role of mathematics, considered from a constructivist perspective, is thus at stake:

> (...) how can something be given to a subject, and how can the subject give something to itself? Here, the critical requirement is that of a constructivist logic which finds its model in mathematics. The critique is empirical when, having situated ourselves in a purely immanent point of view, which makes possible a description whose rule is found in determinable hypotheses and whose model is found in physics, we ask: how is the subject constituted in the given? The construction of the given makes room for the constitution of the subject. The given is no longer given to a subject; rather, the subject constitutes itself in the given. [7]

Becoming is viewed as the creative principle arising from the position of a problem in terms of a constellation of differential operators heterogeneous among themselves.

This phase of the plastic composition of differentials puts in place the problematic and intensive dimension of becoming, which can be regarded as a form of plasticity of the virtual.

---

[1] To be precise Deleuze considers another temporality beyond the Khronos and the Aion: the one of Kyros that is the instant of integration of a differential assemblage with an evenemential character. We will consider later on this temporal axis.

Mathematics can then be used as a language to evoke the dynamical becoming of a complex materiality endowed by its substantial consistency as a vital, singular, semiogenetic flow.

Besides this intrinsic motivation, there is also a historical contingent factor that pushes us to elaborate heterogenesis mathematically.

As in Albert Lautman's epistemic view, mathematics is considered a language that is always relative to specific and situated problematic circumstances, in which an important part of invention consists of the formulation of problems. The history of mathematics, from such a perspective, is considered a history of problems, rather than an automatic progress independent from cultural and historical context, as would be the case according to the axiomatic perspective. The work of mathematicians is to envision the entire problematic dimension in an original way [29].

Accordingly, our interest in the question of heterogenesis resides more in the problematizations it renders possible than in the solutions that it offers. In particular, we are interested in problematizing certain contemporary models that are predominant in both the life sciences and human sciences.

Models in life sciences and human sciences, from the cognitive to the social point of view, from the aesthetic to the semiotic aspect, come from a culture of physical science. By this we mean that these models—which, according to the Diltheyian dichotomy, position themselves on the side of the (determinative) sciences of nature against the (interpretative) sciences of the mind—claiming the forms of a triumphant scientificity, take up, often without critical discussion, the categories and principles of "classical" epistemology, and especially, at the level of empirical a priori, in that they consider an invariant and homogeneous distribution of operators.

This nomological use of operators is at the base of contemporary modelling culture: the Navier-Stokes equation for viscous fluids, for example, is the same in all points of space and time. Analogously, Alan Turing's [33] equation of morphogenesis, deeply studied also by René Thom, presents spatial and temporal symmetries. But living and perceptual becoming, insofar as they are brought to light by the question of heterogenesis, are radically different.

4.1.5 Living and Perceptual Mutations

Within the realm of life sciences, a deep problematization of invariances and symmetries and the necessity of evolving possibility spaces has been proposed by Giuseppe Longo [1, 18]. The possibility space is the space where the scientific description and determination of a phenomenon is given. One major aspect of biological evolution is the continual change of the pertinent phase space and the unpredictability of these changes. "To summarize, the mathematical challenge, with regard to current physico-mathematical theories, consists in the non-punctuality of the structural stability of life (extended criticality relative to numerous control parameters) as well as in the difficulty of establishing a fixed landscape (phase space), within which any process would unfold, following geodesics punctuated by critical transitions. In this regard, it is the phase space itself which changes dynamically (a new organ, a species—unexpected observables and parameters—grow during the course of ontogenesis, of phylogenesis): the dynamic can also be found in the very observables and parameters of the ecosystem, a coevolutive framework where the emergence of novelty changes the basic situation" [1].

In cognitive neuroscience, heterogenesis introduces the possibility of studying the virtual conditions for the deployment of enacted and embodied cognition, where the body is at the centre of the process of perception/action. The cinematic and dynamic body allows the development of enacted cognition [21] while the affective and emotional body also allows for the emergence of saliences and pregnancies in embodied cognition ([32] (1981-1990)). Pregnancies in particular play a pivotal role in plasticity and learning, as it's enough for an experience to have a meaning for the situated body for it to produce plastic variations. So, the affective body allows for a plasticity of functional architectures, or the continual change of differential operators and of their possibility spaces. Symmetries present in classical neuromathematical models are deformed and broken by plasticity reinforced by significant experiences, thus putting in place a true metamorphosis of the virtual. The presence of saliency/pregnancy axis opens onto expressive spaces already in the first experience of perception, putting in place a protosemitic space with the classical expression/ content biplanarity well before any existence of a sign. Finally, the composition of embodied and extended assemblages allows for the introduction of imaginative dimensions and insight experience in 4E cognition.

4.1.6 Negative Results

Heterogenesis can also be thought in line with the negative results that characterize non-positivistic mathematics. Such a mathematics of negative results, instead of determining and restricting the space of possible solutions, opens onto new spaces of freedom. For example, the beautiful result of Poincaré's uncertainty shows the impossibility of predicting the evolution of a system because, even if the initial conditions vary infinitesimally, the trajectories can be infinitely separated. The classic example is the one of the flapping of the butterfly's wings that produces the hurricane. Heisenberg's uncertainty is an even more radical example of a negative solution: here it is shown that the state of a system is indeterminate not due to limitations in the accuracy of a measurement, but for intrinsic reasons due to the non-commutativity of the relevant quantities. That is, it is an insurmountable indeterminacy. Then there is the most beautiful result, that of Gödel's incompleteness theorem, which demonstrates that mathematical practice cannot be reduced to a formal logical system, but is in constant need of problematic externalities. With a series of theorems, Gödel sweeps away Hilbert's logicist positivist project and opens up a problematic dimension of mathematics, wherein the invention of the problem is no less important than its resolution. A negative result could also concern heterogenesis. All the work of the physicist consists in seeking the virtual, that is the differential which lies behind the becoming of forms. The physicist can do it because the virtual of physical forms is somehow fixed. It is even more difficult to search for the virtual of ecological or semiotic systems which, according to the structuralists, are fixed except for the change of parameters within a space of possibility that is given a priori. Looking at it closely, however, the virtual of living and social systems is not fixed, nor is its space of possibility even given a priori. This leaves us with an important question: if the virtual changes punctually in both space and time, what about the practice of making models? Is it still possible and, if so, under what conditions? Once again, the positivist's point of view arrives at an impasse; this impasse, however, can itself be seen to consist in the opening of new spaces of possibility.

4.1.7 Against Homogenesis

As Franck Jedrzejewski outlines in his article "Hétérogenèse et consistance ontologique chez Deleuze et Guattari" [16]:

> Most of the time, heterogenesis is constructed in reaction to a latent homogenesis: here a globalization of capitalist societies, which decline elsewhere, and for example, in the disappearance of auteur cinema or the extinction of non-European languages or cultures. The same process occurs in the therapist."[2]
>
> In this spirit, heterogenesis, continues Guattari, is the war machine against "the capitalistic homogenesis of the generalized equivalent, that leads to the fact that all values are equal, all appropriative Territories are related to the same economic yardstick of power, and that all existential wealth falls under the thumb of the value d exchange. ([12]: 82-83).

Clearly then different kinds of homogeneity are then at stake: there is the homogeneity of the Turing machine (that is the homogeneity of logical algorithms), as well as the homogeneity of the generalized equivalent (that is the homogeneity of money), and the dictatorship of profit as a unique value.

Baudrillard (1981) concludes that these three homogeneities are different expression of the one of information; and A. Rouvroy shows how, under the sign of information, the plane of immanence of singular forms is subjected to algorithmic control and extractivist dynamics. [27]

The purpose of differential heterogenesis can thus be understood to reside in dynamic becoming free from any form of unitary and totalizing symmetry, and in developing thereby forms, action, and thought by means of dispositives of proliferation, juxtaposition, and disjunction. In stark opposition to axiomatization, it aims to problematize morphodynamical becoming from a constructivist perspective.

---

[2] La plupart du temps, l'hétérogenèse se construit en réaction à une homogenèse latente: ici, une mondialisation ou une globalisation des sociétés capitalistes, qui se déclinent ailleurs, et par exemple, en la disparition du cinéma d'auteur ou l'extinction de langues ou de cultures extra-européennes. Le même processus se produit chez le thérapeute.

4.2 Sub-Riemannian Geometric Multiplicity

4.2.1 Beyond Riemannian Geometry

In A Thousand Plateaus, the concept of multiplicity was introduced on the base of Riemannian manifolds, due to the combination of spatial heterogeneity and patchy organization that these offer. Contemporary mathematical research, however, foresees the development of geometries that are even more heterogeneous. Starting from the work of a number of mathematicians in the '60, it became clear that it was possible to further remove some of the constraints imposed by the Riemannian differential calculus. Mathematicians like Hörmander [15], Rothschild & Stein [26], Bony [3], Nagel [20] introduced a more general notion of manifold, called sub-Riemannian, which allows to define a differential multiplicity in a more heterogeneous way.

It is interesting to note that these studies have been developed in the same time span in which the two main texts "Difference and Repetition" and "A Thousand Plateaus" were written, even if these respective developments were evidently independent of one another.

On the side of spatial dynamics, general second-order partial differential equations with non-negative characteristic form have appeared in literature since the early 1900s. They were first studied by [24], who called them elliptic-parabolic equations and proved the celebrated weak maximum principle for their solutions.

The interest in this type of equations in application fields was then established by A. D. Fokker, M. Planck, and A. N. Kolmogorov, who demonstrated that these kinds of equations arise in the mathematical modelling of theoretical physics and of diffusion processes [10, 17, 25]. These studies were motivated by both pure mathematics and the physics of gases.

One of the first works in which the name sub-Riemannian geometry was introduced, establishing the subject as a part of geometry studies, is the paper of Strichartz [31]. In the new geometry, called sub-Riemannian the allowed directions of dynamic propagation are described by vector fields that differ from point to point, giving rise to continuously changing planes of propagation, which are referred to as admissible planes.

A sub-Riemannian manifold is a differential manifold M where at every point just a subspace of the tangent space, called the admissible tangent space, defines the geometry of the space around the point. Even though the tangent space has the same dimension at every point, the admissible tangent space will have dimension changing point to point.

This space of admissible directions introduces a level of heterogeneity unimaginable in Riemannian manifolds, at the prize, however, of introducing considerable difficulties in stating the connectivity between points of the manifold.

In fact, the impossibility of moving in certain forbidden directions makes it difficult to connect all points of the space with admissible curves (see 4.2.5). But at the same time, it opens onto the possibility of introducing cuts and isolated regions on a manifold, which encourages the constitution of disconnected regions. Sub-

Riemannian manifolds thus make it possible to deal with very heterogeneous geometries and also to enlarge the possibility of composing spaces in a fragmented and disconnected way.

The connectivity problem is non-trivial in the sub-Riemannian setting, as pertains to both the question of connectivity between points, due to the very heterogeneous nature of this geometry, and the introduction of disconnected regions and foliated spaces. We will see in the following paragraphs that under the so-called Hörmander condition we obtain completely connected space, while the Frobenius condition induces fragmented and disconnected regions.

4.2.2 *Tangent Space and Admissible Tangent Space*

In a Riemannian manifold, each point can move in any direction. This means that for every point p the tangent space at the point p coincides with the set of admissible directions of the displacements on the manifold from that point. This constraint can be removed, introducing at every point p an admissible tangent space $AT_p(M)$, which is a subset of the $T_p(M)$ at that point. If the tangent space $T_p(M)$ has dimension n, the admissible tangent space $AT_p(M)$ will have dimension $m < n$.

The same construction is repeated for every point. In analogy with the definition (3.7) we will call admissible tangent bundle the union of admissible tangent spaces at every point

$$AT(M) = \cup_{p \in M} AT_p(M).$$

In Fig. 4.1 we depicted tangent and admissible tangent bundles. On the left the surface depicted in grey is a 2D manifold M. The tangent space at every point has dimension two (the grey squares), while AT_p has dimension one (line in red) in every point of the manifold M. On the right the manifold M has dimension 3 and at every point the AT_p has dimension 2 (in red).

The basis of T_p will be composed by n vectors, while the set AT_p will have a basis $(\mathbf{v}_{1,p}, \mathbf{v}_{2,p}, \dots)$, composed by m vectors.

The directional derivatives associated to the fixed basis of AT_p at the point p will be denoted $(X_{1,p}, \dots, X_{m,p})$.

Note that the family of these vectors, considered as functions of the variable p, define m vector fields $(X_1, \dots, X_m)$.

For every i the vector field X_i is defined on M with values in the admissible tangent space to M at the point p.

The complete list of these directional derivatives defines a non-standard gradient,

$$\nabla_p = (X_{1,p}, \dots, X_{m,p}).$$

Note that this gradient has m components in a n dimensional manifold. Hence, it is degenerate from a Riemannian point of view.

It is clear that, if we assign the distribution $AT(M)$ at every point, we can choose a basis at every point. Vice versa an admissible tangent plane at every point can be assigned by means of the choice of a family of vector fields.

The simplest geometric structures in which this can happen are Lie groups. The manifold M is a Lie group if it is equipped with a group law, meaning that for every couple of elements $p, p_0 \in M$ it is defined a composition $p \circ p_0$ and it belongs to M. Moreover the composition law satisfies the associative law, has an identity element and as well as an inverse. Besides, the group law is differentiable with respect to the differential structure of the manifold.

For a fixed point p_0 we can define the group translation which is simply the transformation obtained by applying the composition law to every point. Every point p is moved to $p_0 \circ p$.

If we fix a basis of the admissible Tangent plane AT at the origin,

$$X_{1|_0} = \partial_{x_1}, \dots X_{m|_0} = \partial_{x_m}$$

we can obtain a basis of the admissible tangent space AT_p at every other point p via group law translations. The resulting vector fields will be denoted by

$$X_{1|_p}, \dots X_{m|_p}. \tag{4.1}$$

This ensures that the admissible tangent plane has the same dimension at every point.

A typical example of a Lie group is the group of rigid motions of the plane, that is the group denoted by $SE(2)$ of translations and rotations. If we compose two transformations, each one being a rotation followed by a translation, we obtain again a translation followed by a rotation. Elements of the group are parametrized by a vector (x, y), associated to the translation and an angle θ, associated to the rotation. The whole set can be identified in the space $R^2 \times S^1$ with elements (x, y, θ) with the composition law induced by the composition of rigid motions. A choice of generators of the AT at every point, obtained by translation of the canonical basis at the origin, is

$$X_1 = \cos(\theta)\partial_x + \sin(\theta)\partial_y, \quad X_2 = \partial_\theta. \tag{4.2}$$

If we choose $AT = span(X_1, X_2)$ as an admissible tangent plane at every point, we get the distribution of planes depicted in Fig. 4.1 right, where every plane is obtained by rotation and translation of a fixed one.

4.2.3 Sub-Riemannian Manifolds and Vector Fields

We are interested now in more heterogeneous geometries, in which the set of allowed direction of motions as well as the dimension of AT_p can change from point to point.

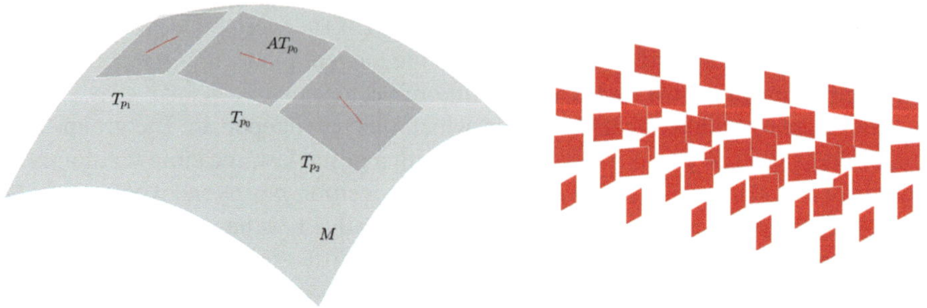

Fig. 4.1 Left: A 2D manifold M with admissible tangent space AT_p of dimension 1 (in red) and the full tangent space T_p of dimension 2 (in grey). Right: a 3D manifold M with admissible tangent space AT_p of dimension 2 at every point

To say that the dimension of AT_p can change from a point to another means that it can be a line at one point and a plane at an other point. Its dimension will be denoted m_p, which is now depending on the point p. Since the dimension of space as well as its generators are locally defined, we can say that "the space change of nature" from a point to an other.

The very heterogeneous structure of the tangent space can induce a variety of directions of propagation of the flows, which change from point to point.

As before at every point p, the set AT_p will be spanned by m_p vector fields $(X_{1,p}, \ldots, X_{m_p,p})$, but this time m_p will change from one point to the other.

A choice of vector fields with different dimension point to point is, for example,

$$X_1 = \partial_x, \ X_2 = x\partial_y.$$

In fact, if $x = 0$ we have an unique vector $X_1 = \partial_x$, so that the dimension of AT_0 is 1. If x is different from 0, then X_1 and X_2 are linearly independent and the dimension of AT_0 is 2 (see Fig. 4.2).

4.2.4 Non-Commuting Vector Fields and Uncertainty Principle

Let us note that on the set of vector fields we can not only apply the operation of addition, but also consider a different operation called commutator, which measures the degree of non-commutativity of the two vector fields.

Indeed, we have identified a vector field X_i with the directional derivative in its direction, hence we can apply two derivatives in sequence, and define a differential operation on AT, called commutator or bracket. Precisely, if X_1 and X_2 are directional

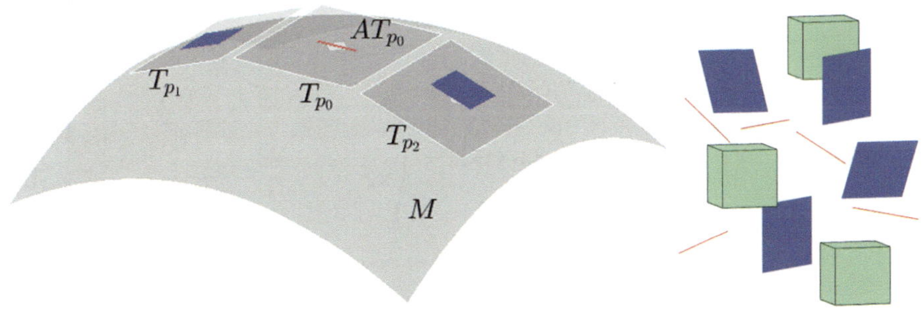

Fig. 4.2 At every point of M we have a tangent plane T_p, and an admissible tangent plane AT_p. The dimension of the admissible tangent plane can be different from a point to the other. On the left a 2-dimensional manifold is visualized, with AT_p of dimension 1 or 2. On the right a 3-dimensional manifold is depicted with AT_p of dimension 1, 2, or 3

derivation operators, also $X_1X_2 - X_2X_1$ is a directional derivative, and it will be denoted

$$[X_1, X_2] = X_1X_2 - X_2X_1.$$

Note that if $[X_1, X_2] = 0$, then $X_1X_2 = X_2X_1$, so that the vector fields X_1 and X_2 commute. For this reason, if $[X_1, X_2]$ is different from 0, it can be considered as a measure of the non-commutativity of the vector fields. An intuitive way to understand this non-commutativity condition is to integrate the vector fields X_i, and evaluate the commutator on curves. Let's consider the vector fields defined in (4.2) and an integral curve of the vector field X_2 with starting point $p_0 = (x_0, y_0, \theta_0)$, that is a rotation around p_0. It will reach a point $p_1 = (x_1, y_1\theta_1)$ after an interval of time of length T. Then we follow an integral curve of the vector field X_1, with starting point $p_1 = (x_1, y_1\theta_1)$. This means that we translate the point $p_1 = (x_1, y_1\theta_1)$ and reach after an interval of the same length T the point $p_2 = (x_2, y_2, \theta_2)$. In this case we first applied the rotation associated to X_2, and then applied the translation associated to X_1. We could apply the same transformations in the reversed order: if we start from the same point p_0, first apply an integral curve of the vector field X_1 and then an integral curve of the vector field X_2. In this way we reach a different point p_3 (see Fig. 4.3). It can be proved that the displacement between p_0 and p_3 has the direction of the commutator $[X_1, X_2]$.

The non-commutativity of these vector fields has a consequence also on the possibility to determining with arbitrary accuracy the state of a system. The most classical version of this principle, called Heisenberg uncertainty principle, states the impossibility to measure with arbitrary precision position, x, and momentum, p. Or, better, it states that the more precisely the position of some particle is determined, the less precisely its momentum can be predicted and vice versa. This principle is ultimately based on the fact that the Heisenberg group is a non-commutative group, and its Lie algebra is generated by two non-commutative vector fields. It has recently been proven [11] that a similar uncertainty principle always holds in presence of

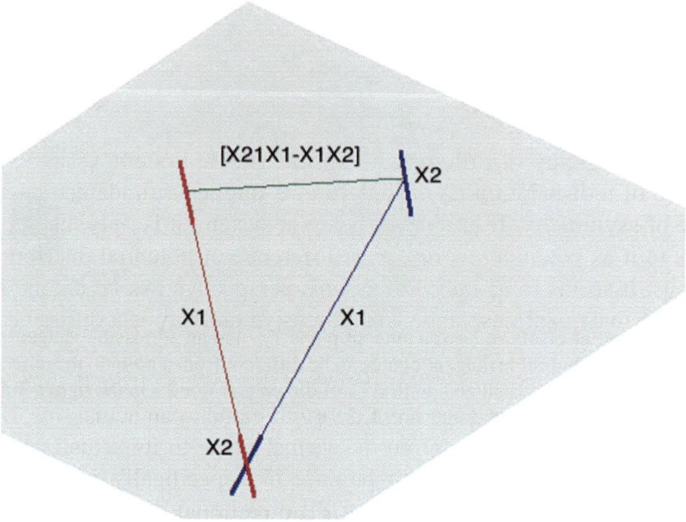

Fig. 4.3 Non-commutative vector fields: if the vector fields $X1$, $X2$ don't commute the composition $X1X2$ is different from $X2X1$. If $X1$ is a translation and $X2$ is a rotation, the application of a translation followed by a rotation (in blu) is different from the application of a rotation followed by a translation (in red). The difference between the two final configurations (in green) is called commutator and is at the origin of the uncertainty principle

non-commuting vector fields: X_1 and X_2, such that $[X_1, X_2] \neq 0$. It is particularly clear if we consider position and orientation: if we consider a point, we can detect its position with arbitrary accuracy, but we have no information to the direction of its motion. If we want to measure the direction of its motion, we need more than one position, and the measure of position will become less precise. *Sub-Riemannian geometry, then, introduces a generalized uncertainty principle that goes well beyond the Heisenberg uncertainty, because it is extended to all non-commutative groups.*

4.2.5 The Sub-Riemannian Metric

In Riemannian geometry we need the definition of norm of a tangent vector (3.4) in order to give a definition of distance. Analogously, in the sub-Riemannian setting, we aim to introduce a similar definition, but on the subspace AT_{p_0}, so that we will define a metric g_{ij} only on the admissible tangent space (Fig. 4.4). We are in presence of a *degeneration of the metric*, from a Riemannian point of view. For every vector v of AT_{p_0} a norm will be defined as in (3.4):

$$||v|| = \sqrt{g_{ij}(p_0)v_i v_j} \tag{4.3}$$

For vectors outside the plane the norm is $+\infty$.

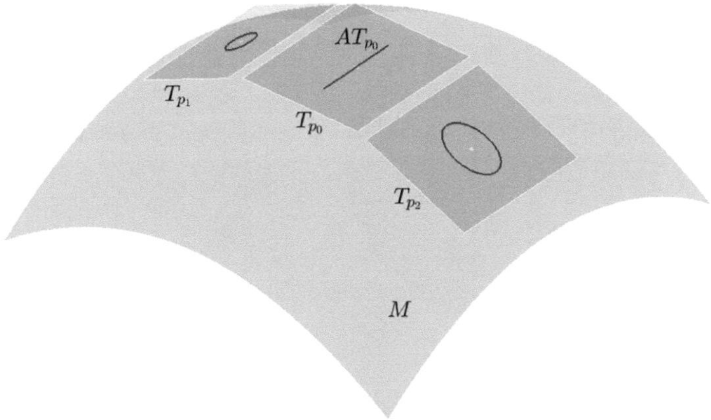

Fig. 4.4 Degenerate metric: In sub-Riemannian geometry the metric tensor is singular since there are forbidden direction of motion

If a curve γ is defined on M and its tangent vector belongs to the admissible tangent space at every point, is called admissible curve.

The distance will be defined as

$$d(p, p_0) = +\infty$$

if there is no admissible curve connecting p, p_0

$$d(p, p_0) = \inf\{l(\gamma) : \gamma \text{ is an admissible curve connecting } p \text{ and } p_0\}.$$

In this setting metric spheres of the distance are very deformed with respects to both Euclidean and Riemannian ones (Fig. 4.5).

4.2.6 *The Connectivity Problem*

4.2.6.1 The Connectivity Problem Under the Hörmander Condition

Now, it is no longer obvious whether or not arbitrary couple of points can be connected by an admissible curve, since there are forbidden directions of motion in the geometry of the tangent space. If, for example, the manifold is the 2-dimensional plane and the admissible tangent space AT_{p_0} consists of vectors we could not move in the vertical direction and we cannot connect points with different vertical positions. Then the connectivity condition would be violated.

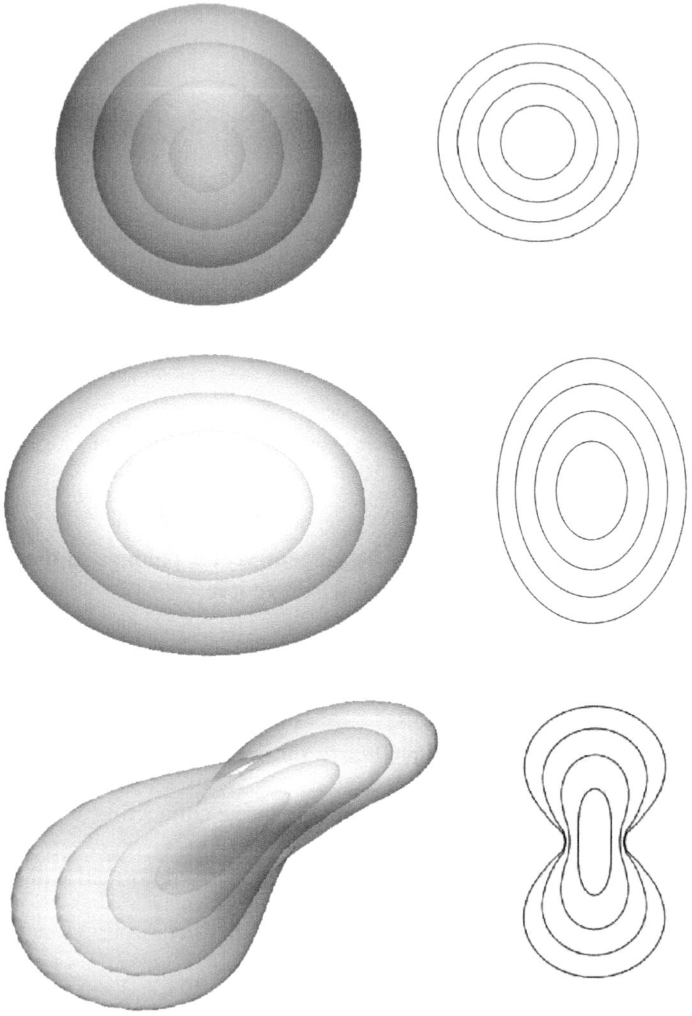

Fig. 4.5 Left: Metric spheres in Euclidean (first row), Riemannian (second row) and Sub-Riemannian (third row). Right: The corresponding projected level sets. Note the resemblance between the sub-Riemannian ball and the twisting of the planes in Fig. 4.1

As a result the connectivity problem is the main problem faced by the sub-Riemannian geometry, starting from the papers [3, 15, 20]. The question at the heart of this work could be expressed as follows. how is it possible to concatenate a pair of points, if some displacements are forbidden? In general, if the horizontal distribution is generated by $X_1, X_2, \ldots, X_m$ we will also introduce vector bundles obtained from AT by applying the commutator (bracket):

$$V_1 = AT, \quad V_2 = [V_1, V_2], \ldots$$

Of course, the vector fields in V_1 will have properties different from the vectors obtained via the bracket. Hence we will assign degree 1 to vector fields belonging to V_1, and we will raise their degree any time we apply the bracket:

$$deg(X) = 1 \text{ if } X \in V_1, \quad deg(X) = j \text{ if } X \in V_j - V_{j-1}.$$

An algebra L_{p_0} is obtained as linear combination of the admissible directional derivatives ∂_{i,p_0} of AT_{p_0} and their commutators. It allows a complete description of the direction of propagation.

Bony [3] proved that the flow associated to integral curves of vector fields in AT will propagate not only along the directions of admissible vector fields ∂_{v_i,p_0}, but also along the direction of the commutators, Hörmander introduced the celebrated condition that the Lie algebra generated by the admissible tangent plane coincides with the tangent plane at every point. This means that the admissible vector fields plus the commutators span the entire space.

Under this condition, any couple of points can be connected by an admissible integral curve by preserving the heterogeneity of connectivity. The Hörmander condition provides the possibility of joining heterogeneous geometries, thereby solving the connectivity problem. However it introduces a generalized uncertainty due to the presence of non-null commutators.

4.2.6.2 The Connectivity Problem Under the Frobenius Condition

If the Hörmander condition is not satisfied, the connectivity property can be lost: there will be couple of points which can be connected with integral curves of the structure, and couple of points which cannot. The space will be organized into connected components. Each component is fully connected, which means that propagation will be possible within each connected component separately, but there will not be propagation from one component to the other.

The shape of the distance ball will change from a point to an other: somewhere it will have the topology of a line, and somewhere else of a sphere. From a purely geometric fact this is unavoidable: Gromov proved that submanifolds of an Hörmander manifold are in general not Hörmander.

This means that there are regions that can escape being connected to other regions that are inconceivable in Riemannian manifolds. Of course, this is a further richness of sub-Riemannian geometry—that is enlarging the dynamical variety (Fig. 4.6).

It's only within each connected component that we can speak about distances, parallel transport, and connections. Space and its connections, however, are not fixed. Rather, the geometry of connections is in continuous dynamic evolution, it is perpetually susceptible to recombination in other heterogeneous structures. A dependence on time can be considered in this case. We can also assume that gradients ∇_{p_0}, which describe the direction of propagation, are not *a priori* fixed, but depend on time, and

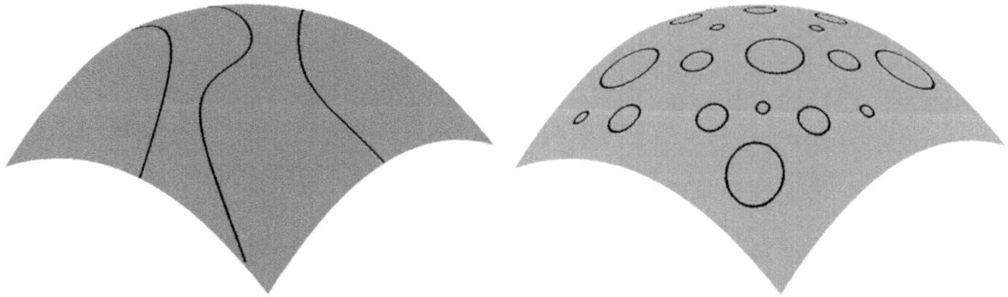

Fig. 4.6 Sub-Riemannian surfaces under the Frobenius condition can be foliated in curves or in isles cutting the space in connected components and introducing the possibility of regions disjoint from the rest of the space

on the dynamic evolution of the solution u. This implies that the vector fields not only constrain the solution, but also, and at the same time, depend on the solution. The structure of tangent planes will be different if the solution has different values.

4.2.7 *Lifting the Geometry of the Space*

4.2.7.1 The Phase Space

A lifting is an increasing of the dimension of the space in such a manner as to allow for an explanation of the dynamics of a phenomenon. The first example of lifting is the usual phase space of mathematical physics where the state of a system is described in terms of the two variables of position and momentum. We have already seen that Poincaré proposed a lifting process in the phase space in order to represent the state of a system (see Figs. 4.1 and 4.2). For Poincaré it is the space of position and momentum that allows for an understanding of the motion of a particle. In this classical setting, the dimension of the phase space is always the double of the dimension of the manifold where the motion takes place. Indeed the state of a system evolving on a manifold M does not depend only on the points of the manifold itself. To the contrary, it depends primarily on the trajectories of the evolution on the manifold. In principle, we would be interested in the space of trajectories, but since the force acting on the system is fixed, then each trajectory, in Hamiltonian coordinates, is uniquely identified by its position and momentum. These are elements of the cotangent bundle: at every point $p \in M$ the cotangent space is defined as the dual space of the tangent space: $T_p^*(M)$, and the dual bundle is defined as

$$T^*(M) = \cup_{p \in M} T_p^*(M).$$

This is why the phase space is able to describe the evolution. In the sub-Riemannian setting the main role of the lifting is to create a common space out of the set of admissible tangent spaces AT. The process can lift homogeneous or heterogeneous AT in a global space, or it can create spaces where homogeneity is local while a global heterogeneity is maintained. The last one is a particularly interesting case because allows heterogenous ATs to communicate in a local space without sharing a global one. These different declinations of the lifting process, from the homogeneous to the heterogeneous, will be faced in the following paragraphs.

4.2.7.2 Lifting of Homogeneous ATs in a Fiber Bundle

Lifting in the cotangent bundle is the most classical example of lifting. Other examples of lifting have been proposed by [4, 9, 14, 23, 30] in order to study the dynamics of neural populations in the visual brain. If the visual stimulus is defined on the 2D retinal plane, the object of the study is the way it is processed by families of cells. Over every point p of the visual plane, a whole fibre of cells is present that is sensible to different values q of the specified feature, and the lifting associates to every point p a couple $\tilde{p} = (p, q)$ of position and feature (see Figs. 4.7 and 4.8). As a consequence, even if the process of vision starts on the 2D visual plane, this latter is not the space wherein brain processing will occur. In fact, the processing takes place on a multidimensional space that associates a variety of visual features to every point of the visual plane.

This space is endowed with a bundle structure more general than a tangent or a cotangent bundle. For every point of the visual plane $p \in M$ there is a whole fibre of features F_p, which can be a vector space or a group. In the simplest cases, the total space M_1 will be defined as the ensemble of fibres

$$M_1 = \cup_{p \in M} F_p.$$

Its elements will be denoted by (p, q), where p is in the basis, and q is the fibre. A natural projection is defined as $\pi(p, q) = p$. In this case, we say that the lifting is global since at every point of the plane we perform the same lifting process.

Anyway the definition of a fiber bundle is more general: M_1 is not necessarily a Cartesian product, but its local behaviour is the one previously described. Indeed A fibre bundle is a (M_1, M, π, F) where $\pi : E \to M$ is a continuous projection. In addition, for every $x \in M$ there is an open neighbourhood $U \subset M$ of x such that

$$\pi^{-1}(U) \text{ is homeomorphic to } \cup_{p \in U} F_p.$$

The purpose of the lifting, then, is to disambiguate the process: at every point p in M we cannot define a unique geometry, since the structure of the space depends on additional parameters q. On the tangent space $T_p M$ it would be impossible to define a metric, since it would depend on the parameter p of the position and q of the fibre

Fig. 4.7 The tangent vectors to level lines of a 2D image are unitary vector fields in the group of rotation and translations. They are lifted in a higher dimensional space (x, y, θ), in order to take into account position and orientation of the level lines

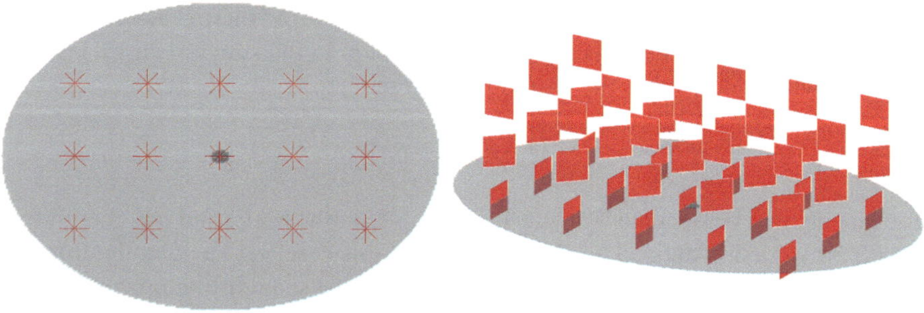

Fig. 4.8 Lifting of the tangent planes $AT(p)$. In this case the tangent planes have the same dimension at every p. Left: The base of the fibre bundle. Right: The lifting of tangent vectors in a higher dimensional space (x, y, θ) is a fibre bundle

(see Fig. 4.7). Enlarging the dimension of the space, also the lifted space M_1 will be a differentiable manifold. Hence at every point (p, q) of M_1 we will have a tangent space $T_{p,q}M_1$ which inherits the geometry from the low dimensional basis M, so that in many examples the dimension of the tangent bundle $AT_{p,q}M_1$ is strictly smaller that the dimension of $T_{p,q}(M_1)$, and M_1 will become a sub-Riemannian manifold.

4.2.7.3 Lifting of Non-Homogeneous ATs

In sub-Riemannian geometry the lifting process is more subtle and consists in the construction of a space with a different Lie algebra defined on every tangent space T_{p_0}. In fact a Hörmander manifold is characterized by the fact that the dimension of the admissible tangent space AT_{p_0} can change from a point to the other.

The main scope of the lifting process of Rothschild-Stein is to show that locally any sub-Riemannian manifold can be approximated via a Lie group, that is, an easy and well-known structure.

From a technical point of view the procedure is based on a disambiguation process of the generating vector fields. Let us consider as an example the manifold $M = R^2$ with the choice of vector fields

$$X_1 = z\partial_y, X_2 = \partial_z,$$

that is visualized in Fig. 4.9. The dimension of the admissible tangent plane is 1 if z vanishes and 2 when z is different from 0. We will lift then with the double scope of obtaining a space where

(1) the admissible tangent space has the same dimension at every point, and
(2) the notion of degree of fields is well defined.

By definition X_1, X_2 will have degree 1, and their commutator will be

$$X_3 = [X_1.X_2] = \partial_y.$$

This vector will have degree 2. Hence the vector ∂_y has at the same time degree 1 and 2. Also note that X_1, X_2, X_3 are linearly independent as vector fields, but if we fix a point p and consider then as vectors, $X_{1|p}, X_{2|p}, X_{3|p}$ will be linearly dependent. We can enlarge the space to clarify the notion of degree. Hence we will call $\tilde{M} = R^3$, and we will denote $\tilde{x} = (x, y, z)$ its points. Then we call

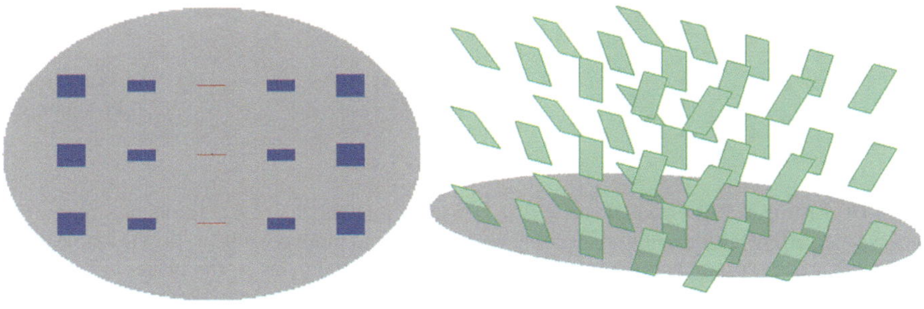

Fig. 4.9 Lifting of the tangent planes $AT_{p_0}(p)$ in the neighbourhood of p_0 generates a fibre bundle, providing the tangent planes to the local phase space. Admissible tangent planes with different dimensions on the base space (left) have the same dimension when lifted (right)

$$\tilde{X}_1 = \partial_x + z\partial_y, \tilde{X}_2 = \partial_z \quad \tilde{X}_3 = [X_1.X_2] = \partial_y.$$

In this way the vector fields $\tilde{X}_1, \tilde{X}_2$ are linearly independent from $\tilde{X}_3$ and we can assign degree 1 to $\tilde{X}_1, \tilde{X}_2$, degree 2 to $\tilde{X}_3$.

4.2.7.4 Local Lifting

The construction we have just seen can be only local and hold for an arbitrary family of vector fields

$$X_1, \ldots X_m,$$

acting on the admissible tangent plane of a manifold M (Fig. 4.10). For a fixed point p_0 there will be a ball B_{p_0} centred in p_0 such that for every point p in B_{p_0} the degree of vector fields in p will be bigger than the degree of the correspondent vectors in p_0. In this neighbourhood of p we will apply the previous lifting procedure but with more delicate considerations, as in the following. First, the vector fields, $X_1, \ldots X_m$ which satisfy Hörmander's condition at some step r, are lifted to some new vector fields

$$\tilde{X}_1, \ldots \tilde{X}_m$$

belonging to the admissible tangent space to a higher dimensional manifold $\tilde{M}$, which satisfy Hörmander's condition at the same step r, and are free up to step r. Second, one proves that there exists a structure of homogeneous group $\tilde{G}$ and a family of left invariant homogeneous vector fields $\tilde{Y}_i$ which locally approximate the $\tilde{X}_i$. In this way it is possible to locally reduce the study of the geometry of the space to the study of a homogeneous left invariant operator. Indeed the lifted space will be a group, so that the admissible tangent space will be locally the same at every point (Fig. 4.11).

Fig. 4.10 Lifting of the tangent planes $AT_{p_0}(p)$ in the neighbour B_{p_0} of p_0. The admissible tangent planes have different dimensions depending on p and it is no more possible to lift the whole patch. In the left image we see that in a smaller neighbourhood of the point p_0 the planes $AT_{p_0}(p)$ can be approximated with the $AT_{p_0}(p)$ in Fig. 4.9. The lifting is now local and performed only in this neighbourhood (right)

Fig. 4.11 Different features of an image are defined by different vector fields in a continuous way. Lifting is heterogeneous in this case and particularly it is local

This passage of heterogeneous vectors to a Lie algebra has to be considered as the core of the process of lifting introduced by Rothschild and Stein.

Notice that this procedure allows for the construction of local common spaces without destroying the heterogeneity of vector fields. This is likewise the case for Leibnizian monads, for which extreme heterogeneity does not prevent the possibility of local resonation. In any case the Rothschild and Stein construction is based on the local fulfilment of Hörmander's condition. If just the Frobenius condition is satisfied then disconnected regions and isles are preserved.

4.2.7.5 General Lifting Process

We are interested in lifting processes including all the possibilities we have described. We will consider manifolds, which have at some point multiple feature selection, and non-unique local admissible tangent space, to be lifted to a fibre bundle. The dimension of the admissible tangent space will be different from one point to another within a neighbourhood of other points, and we will operate a RS lifting in order to enlarge the space and reduce it to a group.

It is clear in this construction that the dimension and the geometry of the lifted space is given by the immanent choice of vector fields on the tangent planes. In this sense, the phase space is not given *a priori*, but it is induced patch by patch, tangent plane by tangent plane, changing point by point on the manifold and in time. Lifting is namely the construction of the possibility space for any dynamics and it is an immanent process far from nomological determinations. We will see in the next chapter that a multiplicity of dynamical operators will populate these heterogeneous spaces.

4.3 Heterogeneous Dynamic Multiplicity

4.3.1 Differential Operators

To this point, we've limited our focus to manifolds and their geometry. It's crucial to recall, however, that the manifold is the space wherein a specific dynamic action takes place. This action, in turn, induces the evolution of forms, to which we must also now turn our attention. Dynamics is classically described by defining a differential operator A. In chapter "Structural morphodynamics" we considered systems of ordinary differential equations represented by $\frac{du(t)}{dt} = A(u(t))$, where the solution $u(t)$ only depends on the time variable, and A is a function of u. Evolutions of this kind deal with phenomena like the motion of particles, planets, orbits, pendulums, oscillators, and so on.

In what follows we will turn our attention instead to systems of evolutive partial differential equations, represented as

$$\frac{\partial u(t, p)}{\partial t} = A(u(t, p)), \tag{4.4}$$

which generalize the previous one, since the solution $u(t, p)$ depends on space and time, and the operator A depends on the partial derivatives of the solution. Evolutive partial differential equations deal with processes like wave propagation, crystal growth, development of biological patterns, evolution of cities, and any kind of individuation and morphogenetic becoming.

In this setting an operator is represented in terms of the spatial gradient ∇, and its iterations of any order ∇^k representing spatial differences of any order.

The expression of an operator A could be thought as a function $a(u)(p)$:

$$A(u)(p) = a(t, p, u(p), \nabla u(p), \ldots, \nabla^k u(p)), \tag{4.5}$$

where a depends on the spatial position p.

It is clear that according to this definition, a is not required to satisfy any smoothness condition, not even to be continuous. Hence it can vary arbitrarily and abruptly from one point to the other, in time and space, with absolutely no criterion.

Even more generally, the differential operator can be stochastic, leading to a solution that is also a stochastic process. In order to define this type of operators, we could consider a probability space (E, P, μ), where E is a space, P is a $\sigma-$algebra on E, and μ is a probability measure. Then a stochastic differential operator is an operator

$$A(u)(\omega, p) = a(t, \omega, p, u(p), \nabla u(p), \ldots, \nabla^k u(p)), \tag{4.6}$$

where $p \in \Omega$, $\omega \in E$.

Expressions (4.5) and (4.6) represent choices of operators at every point, in the spirit of the Deleuzian throw of the dice.

In order to clarify this, we recall the definition of a function. Since a is a function, to every element $(t, \omega, p, u(p), \nabla u(p), \ldots, \nabla^k u(p))$ on the domain it associates a different real value. This is a punctual definition. At each point in space and time, and for all values of the derivatives, we only require that a attains a real value. There is no required relation between the values of a at a point p and at an other point p_0. With no requirement of regularity the function a changes in an arbitrary way from one point to the other. In this sense a function a is the most general example of arbitrarily throwing a different operator at each different point.

With this generality, nothing can be proved from the point of view of mathematical analysis. That is, under these general assumptions we are not able to say anything about the properties of solutions. Results are known only if we make very restrictive assumptions on the operator: it is generally required that the operators satisfy suitable conditions uniformly on the whole set.

If we want to prove the existence or non-existence of a solution, we need to impose a very strong relation between the dynamics and the geometry of the manifold.

For example in the Euclidean setting the gradient is simply the full collection of all the partial derivatives:

$$\nabla = (\partial_1, \ldots, \partial_n).$$

In a Riemannian manifold characterized by a metric (g_{ij}), the Riemannian gradient will be defined as

$$\nabla_g = \left(\sum_j g^{1j} \partial_j, \ldots, \sum_j g^{nj} \partial_j \right).$$

This means that the gradient is composed by n directional derivatives, weighted by the coefficients of the metric. In a sub-Riemannian manifold of topological dimension n, the geometry is defined in terms of m vector fields $X_1, \ldots X_m$ which define also the sub-Riemannian gradient:

$$\nabla_S = (X_1, \ldots X_m). \tag{4.7}$$

The sub-Riemannian gradient has only m components, with $m < n$, and it is totally degenerate. This means that the solutions will mainly propagate along the directions of the horizontal vector fields. The dynamic will also propagate along the direction of the commutators, but at lower speed. We see then that operators contain the geometry of the space and they can be considered primary with respect to the geometry.

4.3.2 Sub-Riemannian Flows. Homogeneous Operators in an Heterogeneous Geometry

Let us consider a sub-Riemannian manifold, with admissible tangent bundle $AT = span(X_1, \dots X_m)$, and a metric g defined only on the admissible tangent plane (see expression (4.3)). The sub-Riemannian gradient is defined in the expression (4.7). Since the sub-Riemannian geometry can change continuously at every point we will have a different gradient

$$\nabla_{S|p_0} = (X_{1|p_0}, \dots X_{m|p_0}).$$

The works of [26] and Jerison (1986) as well as large part of the subsequent literature in this field , consider second-order operators and the corresponding flows

$$A_{SR} = \sum_{i=1}^{m} a_{ij} X_{ij}^2 + \sum_{i=1}^{m} b_i X_i^2 + c X_0, \tag{4.8}$$

where a_{ij} is the inverse of the metric g_{ij}, and b an arbitrary vector. When $c = 0$, the operator is called uniformly subelliptic and its flow is a sub-Riemannian diffusion, when c does not vanish, we have a Fokker Planck operator with the corresponding advection-diffusion flow. The evolution equation (4.4) applied to this operator defines the sub-Riemannian flow:

$$\frac{\partial u(t, p)}{\partial t} = A_{SR}(u(t, p)).$$

4.3.3 Heterogeneous Operators

We are interested in dynamics that can vary in space and time.

More specifically we are interested in a level of heterogeneity where operators can be different in type and order at different spatio-temporal points. For example, we could have an advective operator at one point and a diffusive behaviour at other points (see, for example, [19]).

More generally we can consider a constellation of operators with totally different dynamics at each point: A discrete set of differential operators A_{p_0,t_0} each one defined in a neighbourhood U_{p_0,t_0} of a different point p_0, starting its existence from a different instant of time t_0:

$$A_{p_0,t_0}(u)(p) = a_{p_0,t_0}(p, u(p), \nabla_{p_0} u(p), \nabla^2_{p_0} u(p)).$$

Note that the domain $U_{p_0,t_0} = U_{p_0,t_0}(t)$ can depend on time $t > t_0$, as for example in the dynamic of erosion, or flame propagation or free boundary propagation. Of course, the operator and its domain can have an extinction time $T > t_0$, after which the dynamics is no longer defined, as frequently happens, for example, in free boundary problems: think for example to a frozen region, in the surface of a lake, which is melting, so that its boundary is evolving.

Let's try to understand under which conditions and in which sense this family of operators A_{p_0,t_0} define an operator A as the assemblage of the previous ones.

We have to consider a phenomenon with many self-contained parts, each one evolving separately, with their own internal organization. If two or more of these parts start interacting at some point, their internal organization and connection will be completely reorganized, to take into account the new interaction. We will be precisely interested in this process even more than in the final operator A itself.

In particular we do not fix a priori a Riemannian or sub-Riemannian metric. We choose a manifold and a family of operators on it. The metric and the geometry will be induced by the choice of the operator. In this manner, operators become primary and thus define dimensions and qualities of the space, breaking the Kantian a-priori of the primacy of the space in any morphodynamics.

4.3.4 *Operators as Shapes*

In our approach the operator is primary, and we will deduce the geometry of the space from its properties. In order to do so, it could be convenient to identify the operator with a function, or a shape.

Two approaches are possible: using the fundamental solution or a Gaussian transform.

4.3.4.1 The Fundamental Solution

For linear operators, it is useful to introduce the inverse operator A^{-1}. It contains the same information as A, in particular, it contains the geometry of the space, in the sense that its level sets correspond to a family of metric spheres (see Fig. 4.12).

In addition A^{-1} is an integral operator, so it is represented in terms of a fundamental solution Γ. Integral operators are more stable than differential operators, (since they

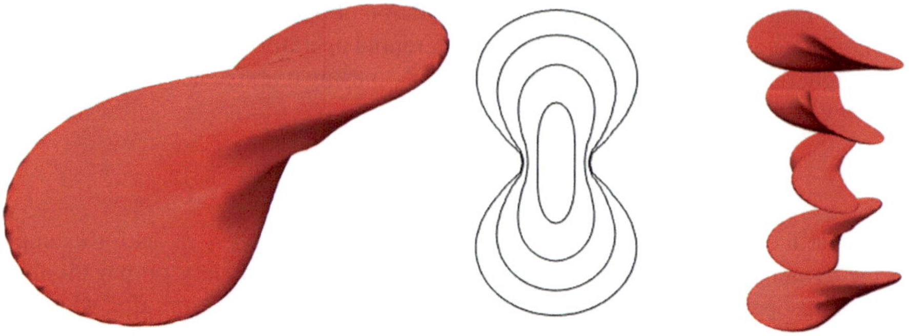

Fig. 4.12 Left: Fundamental solution of hypoelliptic diffusion. Note that its level set corresponds to metric spheres. Centre: Level sets of the solution. Right: The fundamental solution lifted on a fibre of rotations

gain regularity, while differential operators lose regularity). Hence it is, in some cases, more convenient to work with the fundamental solution. The interest of this solution Γ is that it allows to represent any other solution. Indeed the solutions, with forcing term f can be simply expressed as

$$u(x) = \int \Gamma(x, y) f(y) dy. \tag{4.9}$$

This provides, when available, a simple and direct instrument to find solutions.

The problem of existence of a fundamental solution is totally solved in the case of Lie groups. In this case the operator A is left invariant with respect to the group law. As a consequence its fundamental solution can be obtained by a fixed one by application of the group law. As an example we can consider $R^2 \times S^1$ with the group law of rotation and translation, and the sub-Riemannian metric induced by the choice of vector fields defined in (4.2). A fundamental solution of the subelliptic Laplacian (or subelliptic flow at fixed time) is depicted in Fig. 4.12 on the left, and a whole fiber of fundamental solutions on the right. By comparison with Fig. 4.5 we see that the level set of the fundamental solution coincides with the ball of the metric. On the other side, the fundamental solutions are obtained by rotation of the fixed one, as the plane of the structure, depicted in Fig. 4.1 on the right.

The fundamental solution of the Fokker Plank flow has an analogous structure (see Fig. 4.13), since the fundamental solution at every point can be obtained by rotation and translation from a fixed one.

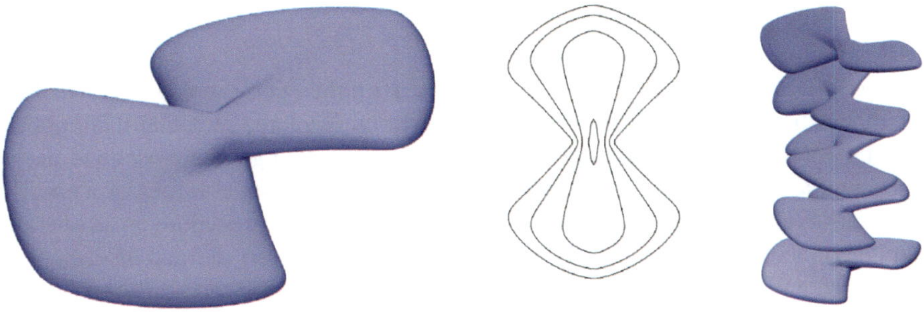

Fig. 4.13 Left: Fundamental solution of the Fokker Planck operator. This fundamental solution contains information about both geometry and dynamics. Centre: Level sets of the solution. Right: The fundamental solution lifted on a fibre of rotations

4.3.4.2 Gaussian Transform

A simpler and more general way to associate a shape to an operator is to consider the Gaussian transform. Precisely applying the operator to the Gaussian function $G_\sigma(x)$, we obtain a function $A_{p_0}(G_\sigma)$, which can be considered a shape and can be also be considered a good approximation of the operator itself. Indeed its convolution with an arbitrary test function ψ tends to the operator applied to ψ as σ tends to 0. Hence for σ sufficiently small $A_{p_0}(G_\sigma)$ can replace A_{p_0}. In addition it is a function, so that we associate to any differential operator a function $A_{p_0}(G_\sigma)$, which can be identified with the density of a measure, which makes it possible to use the language of functional analysis or optimal transport for any operator A_{p_0} (Fig. 4.14).

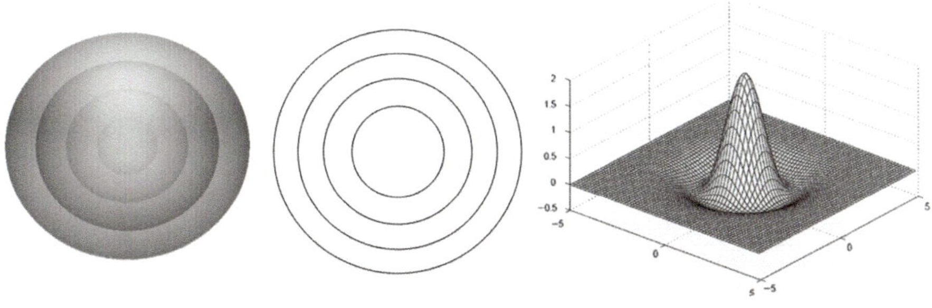

Fig. 4.14 Three geometric representations of the Laplace operator. Left: Its fundamental solution. Centre: Level sets of the fundamental solution. Right: Its Gaussian transform

4.3.5 Lifting of Operators

Representing operators as shapes allow us to clarify the operatorial analogous of the geometric lifting described in the previous sections. We have seen that it is possible to lift a geometry defined on the 2D the retinal plane, to a geometry defined on the fiber bundle over the 2D plane by the selection of all possible values of a specific feature at every point. It is possible to perform an analogous procedure on an operator defined on the 2D space: it will be lifted to a fiber of operators defined on each point leading to a fiber bundle of operators (Fig. 4.15).

There is a strict relation between the geometry of a manifold and the dynamic evolving on the same manifold. If the dynamic is expressed via a PDE, then the metric can be deduced from the shape of fundamental solutions. In Fig. 4.16 we see the strict relation between the two fibre bundles in the case of the rotations and translations group: the two structures present the same tilting feature.

4.3.5.1 Sub-Riemannian Fundamental Solution

Detecting the fundamental solution of a general sub-Riemannian operator is a delicate construction, which was introduced by Rothschild and Stein using a lifting and approximation process. In this case, the geometry of the space changes from one point to the other and can be locally approximated by different Lie groups. As we recalled in the previous section, the manifold M can be lifted to a higher dimensional Lie group $\tilde{M}$, which contains as a subgroup all the local groups, allowing interaction.

If the manifold M has admissible tangent bundle $AT(M) = span(X_1, \dots X_m)$, defined in a neighbourhood U_{p_0} of a point p_0, the lifting process defines a higher dimensional manifold $\tilde{M}$, and lifted vector fields $\tilde{X}_1, \dots \tilde{X}_m$.

The correspondent gradient will become

$$\tilde{\nabla}_{S,|p_0} = (\tilde{X}_{1,p_0}, \dots \tilde{X}_{m,p_0}).$$

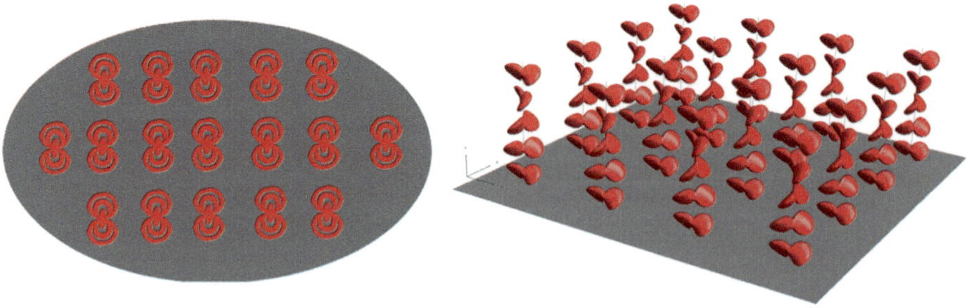

Fig. 4.15 The operatorial analogous of process visualized in Fig. 4.8. Left: The base of the fibre bundle. Right: The lifting of tangent vectors in a higher dimensional space fibre bundle of operators

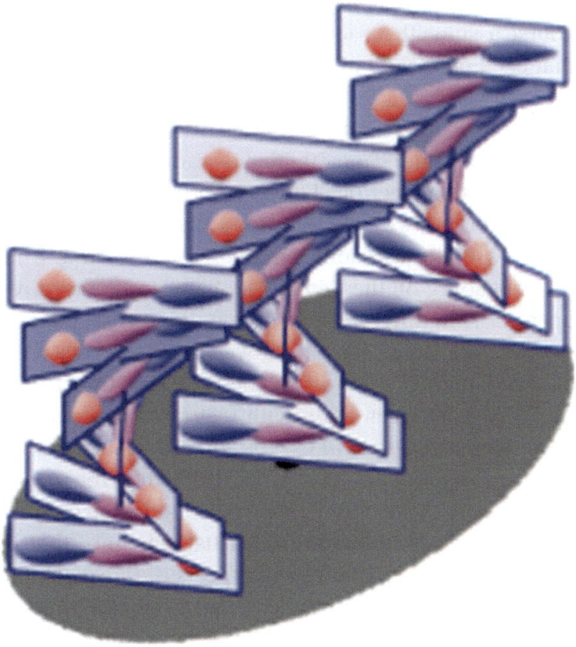

Fig. 4.16 Geometric and dynamic operator bundles: On a geometric bundle of rotation-translation a bundle of fundamental solutions is visualized

The operator A is consequently lifted to an operator $\tilde{A}$ defined on the lifted geometry. Its restriction $\tilde{A}_{p_0}$ are now acting on the same space, and continuously varying from one point to the other. Each of them is invertible, with inverse $\tilde{A}_{p_0}^{-1}$. As a consequence their inverse operators $\tilde{A}_{p_0}^{-1}$ act on the lifted space and continuously depend on the parameter p_0. This means that the inverse operators $\tilde{A}_{p_0}^{-1}$, are different from one point to the other but can be considered infinitesimally adjacent operators, and deformations of the same operators. Exploiting this property with a parametrix method, Rothschild and Stein were able to construct the operator $\tilde{A}^{-1}$ as an envelope of the given operators $\tilde{A}_{p_0}^{-1}$. And to find in this way the fundamental solution of the operator A by projecting back on the initial lower dimensional space.

Up to now the geometry we've considered has been heterogeneous while the operators have been homogeneous. we will now consider families of heterogeneous operators in heterogenous geometries.

4.4 Geometric and Dynamic Assemblages

At this point we know how to deal with heterogeneous distributions of operators. We will now focus on the construction of an assemblage between a family of operators A_{p_0,t_0}, each one operating on regular functions defined on a different local set $U_{p_0,t_0}(t)$, depending on time, which we will assume to be connected. We can of course assume that each operator A_{p_0,t_0} has been lifted independently, with the result that each ends up with a different fibre bundle structure, with different dimensions on each patch U_{p_0,t_0}.

Now our aim will be to conjoin different local multiplicities, each of which entails its own particular geometry and dynamics, remembering that "Rather, an assemblage establishes connections between certain multiplicities" ([6] (1980)).

To proceed, we need to introduce an analogous instrument at the operatorial level, which will allow to define an operator A starting from the local expression A_{p_0}.

Depending on the explicit problem to be faced, we will be able to construct the operator A in different ways. We will interpret the operators A_{p_0} as points in a space of operators. To each operator we will associate a local measure, which can be either the fundamental solution, or the Gaussian transform of the operator. Then, we can find a path between them with instruments of transport theory. In order to conjoin more operators, we will exploit the possibility of finding a surface that interpolates between them. This idea of dynamic heterogeneous concatenation is implicit in the concept of machine elaborated by Deleuze and Guattari. Notice that we deal always with constellations of virtual elements, that is differential operators and not actualized objects. Interesting aspects of this machinic concatenation are introduced and commented in the article "L'hétérogenèse machinique" [13].

As a result of this first process, we can assume that any pair of elements of the multiplicity are already defined on domains of different dimension and have developed its own internal geometry.

The construction of the assemblage is performed in three steps. First, we apply the lifting to the union of all considered operators of the multiplicity, so that they all act on a space of the same dimension. Then we extend each operator of the multiplicity, so that they act on functions defined on the same set. Finally, we define the assemblage operator in the lifted space.

4.4.1 Dynamic and Geometrical Lifting of a Multiplicity of Operators

If we assign two operators A_{p_0,t_0} and A_{p_1,t_1}, they can be the result of a lifting process. Hence their associated neighbourhoods U_{p_0,t_0} and U_{p_1,t_1} can have different dimensions, each of which we will lift to a higher dimensional set, in order to allow intersection. Indeed two operators can be connected only if the lifted domains have a non-vanishing intersection at some time $t_{0,1} > max(t_0, t_1)$.

The admissible tangent associated to the two operators is generated, respectively, by the vector fields

$$\nabla_{p_0,t_0} = (X_{p_0,1}, ..., X_{p_0,m_0}) \text{ and } \nabla_{p_1,t_1} = (X_{p_1,1}, ..., X_{p_1,m_01}).$$

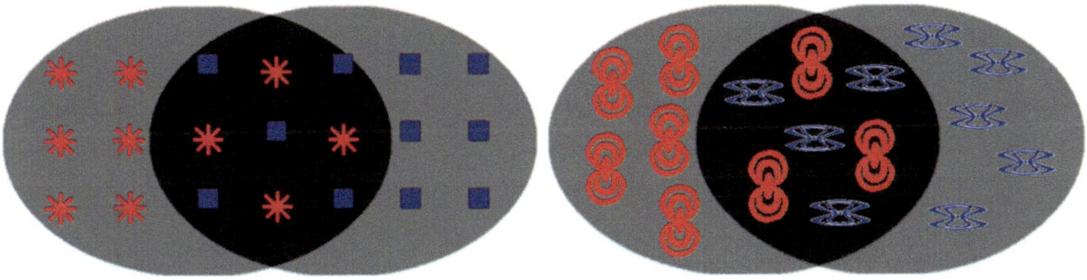

Fig. 4.17 Intersection of two regions. Left: The geometry. In the first regione the admissible tangent plane is generated by vectors fields X_i, in the second one by vectors fields Y_i. As a result in the intersection we have the whole set X_i, Y_i. Right: The correspondent operators

In the intersection of patches $U_{p_{0,1},t_{0,1}} = U_{p_0,t_0} \cap U_{p_1,t_1}$ the generators are the set of vector fields $X_{p_0,i}$, $X_{p_1,j}$ that collect all the directional derivatives of the two different gradients in a new gradient

$$\nabla_{p_0,p_1,t_{0,1}} = (\nabla_{p_0}, \nabla_{p_1})$$

defined for $t > t_{0,1}$ (see Fig. 4.17).

In the intersection, we merged two families of vector fields and a new lifting process is applied to the domains in order to relate them. As a result the intersection of patch will be lifted to new domains $\tilde{U}_{p_i,t_i}$, the vector fields will be lifted to vector fields $\tilde{\nabla}_{p_i,t_i}$ that coincide in the intersection of patches with the lifting of the operator $\tilde{\nabla}_{p_0,p_1,t_{0,1}}$ (Fig. 4.18 first and second row).

This process clarifies how to merge the geometry, but it does not face the problem of the assemblage of operators, since we still have two operators $\tilde{A}_{p_i,t_i}$, with $t = 0, 1$. The lifting procedure is not unique and it is defined only on the basis of the possibility of differential constraints to create new spaces and new operators.

Also recall that the two operators A_{p_0,t_0} and A_{p_1,t_1}, which have been evolved independently up to time $t_{0,1}$, can themself be the result of previous lifting processes.

In any case, the lifting $\tilde{\nabla}_{p_0,p_1}$ contains commutators that did not exist in each of the lifted operators separately: the interaction is therefore much more than the simple union of the collected vector fields. Commutators interpret in a formal way the differences of differences, which is a crucial feature in the Deleuzian construction of assemblages.

4.4.2 Extension of the Operator Via Partition of the Unit

With the previous procedure, we end up with two operators $\tilde{A}_{p_0,t_0}$, and $\tilde{A}_{p_1,t_1}$ defined on different neighbourhoods, but with non-empty intersection. However, we can extend each of them to the whole space assigning value 0 outside its domain $\tilde{U}_{p_i}$. This can be accomplished by multiplying each of them by a function ϕ_i that has value 1 around the point p_i and 0 outside the set $\tilde{U}_{p_i}$. If we also normalize the two functions in such a way that their sum is identically 1, the couple (ϕ_1, ϕ_2) is called a partition of unit.

4.4.3 Heterogeneous Assemblage

Thanks to this extension, the two operators act on the same set of functions, allowing heterogeneous assemblage, which is a deformation of an operator in the other. This process of assemblage can be expressed as a linear combination of the two lifted operators (or a more complex operation involving both of them): &

$$\tilde{A}_{p_0, p_1, t_{0,1}} = \phi_0 \tilde{A}_{p_0, t_0} + \phi_1 \tilde{A}_{p_1, t_1}.$$

Since ϕ_0 takes value 1 around the point p_0 and 0 far from it, the assemblage coincides with $\tilde{A}_{p_0,t_0}$ in a neighbourhood of the point (p_0, t_0). Analogously, since ϕ_1 takes value 1 on at the point p_1, the assemblage coincides with $\tilde{A}_{p_1,t_1}$ in a neighbourhood of the point (p_1, t_1). The resulting operator is then a smooth transformation of $\tilde{A}_{p_0,t_0}$ into $\tilde{A}_{p_1,t_1}$ (Fig. 4.18, third row). Note that this is just one of the many possible recombinations of operators.

The operator can at this point be re-projected to an operator & $\tilde{A}_{p_0, p_1, t_{0,1}}$ on the substrate space.

More generally, we can define an assemblage between two or more heterogeneous patches, if there exist $k + 1$ points denoted p_0, ..., p_k such that the neighbourhood of each one intersects the following one. In each intersection. the previous process is applied, and a common assemblage is defined.

An inverse of the assemblage operator is the disjunction operator: $(\&)^{-1} A$, which is able to generate two distinct operators A_{p_0,t_0} A_{p_1} starting from an unitary assemblage A. This operator is not unique, since the distinct operators can be generated in different ways.

4.4.4 Curve in the Space of Operators and Metamorphosis of Operators

An assemblage can evolve in time. For example considering a curve in the space of operators. There are, in general, infinite choices for a curve connecting two operators (obtained via the lifting and the extension process).

The first possibility is a linear combination of them:

$$\tilde{A}_\tau = \&\tilde{A}_{p_0, p_1, t_{0,1}, \tau} = \tau \phi_0 \tilde{A}_{p_0, t_0} + (1 - \tau) \phi_1 \tilde{A}_{p_1, t_1},$$

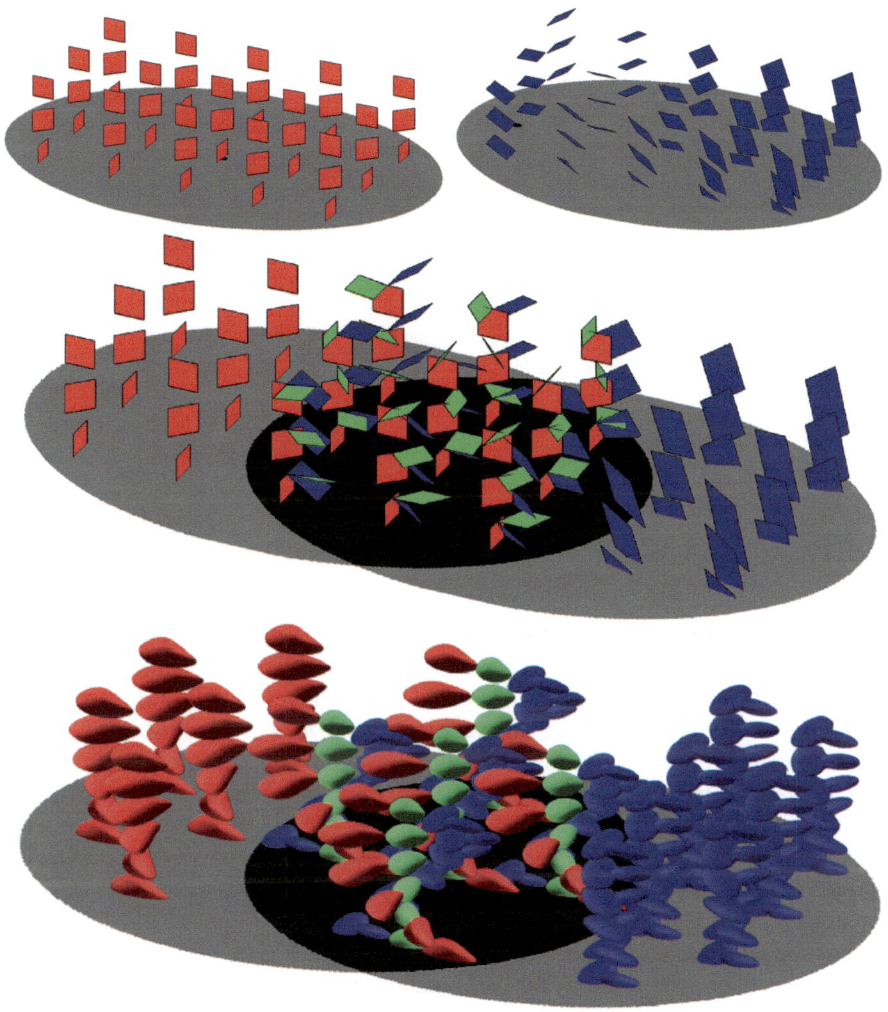

Fig. 4.18 Lifting of the assemblage. Top left: Geometric Lifting of the geometry associated to vector fields $X_i$and an operator A_{p_0}. Top right: Geometric lifting of the operator A_{p_1}, whose gradient is generated by the vector fields Y_i. Centre: Geometric lifting of the assemblage & A_{p_0,p_1}, whose gradient is generated by X_i, Y_i. Bottom: Operatorial lifting of the assemblage & A_{p_0,p_1}, whose gradient is generated by X_i, Y_i. Notice that the lifting of the assemblage is performed by the generators induced by A_{p_0}, the generators induced by A_{p_1} and their commutators (in green). Then the lifting of the assemblage is more than the union of the separated liftings, due to the presence of new commutators (differences of differences in the language of Gilles Deleuze). This assembly of planes indicates the possible directions of flows

Fig. 4.19 Evolution of the operator kernel from a fourth-order diffusion in an Euclidean space to a second-order diffusion in a sub-Riemannian geometry

where A is a curve connecting the two operators A_{p_0,t_0} and A_{p_1,t_1} parametrized by the deformation variable τ. However, the operator also depends on the time variable t, which is the natural time of evolution. The two variables t and τ correspond to the two times of Khronos and Aion. The variable t is the axis of Khronos and describes the propagation of the process in time. The variable τ describes the evolution of the differential constraints in new assemblages. giving rise to new dynamics instant by instant.

For example in Fig. 4.19 it is shown the evolution of the operator kernel from a fourth-order diffusion in an Euclidean space to a second-order diffusion in a sub-Riemannian geometry. Notice that the linear combination is just one illustrative way among the infinite ways to join two different operators. It is this type of? transformation that accounts, for example, for the so-called "rare event" in phylogenesis, which allows for the mutation of living being.

Another evolution of operators that we can consider is one of the gradient systems $x'(t) = -\nabla V(x)$ where the potential V defines a specific kind of Thomian elementary catastrophe. After the introduction of the time axis τ of Aion, these operators can evolve considering the transformation $V_\tau = (1-\tau)V_1 + \tau V_2$, where V_i corresponds to the potentials of elementary catastrophes. In this way, one catastrophe can evolve into another one, as shown for example in Fig. 4.20 where the catastrophe of Fold evolves in the Cusp and finally in the Swallowtail. Analogously, in Fig. 4.21 the evolution is not smooth but scattered.

On the other hand, after applying the Gaussian Transform or considering the fundamental solution, we have identified any operator with a density measure, allowing for the use of curves in function spaces to connect operators. If the operators act on functions defined on a subset U of the whole space, the associated density will

Fig. 4.20 Evolution of Thom's elementary catastrophes with potential $V_\tau = (1-\tau)V_1 + \tau V_2$ from the V_{fold} (first frame) to the V_{cusp} (4th frame) and to the $V_{swallowtail}$ (last frame). The equilibrium manifold $\frac{\partial V(x,p)}{\partial x} = 0$ is visualized

Fig. 4.21 Discontinuous evolution of Thom's elementary catastrophes: The evolution is scattered in such a way that catastrophes suddenly change in the sequence V_{fold}, $V_{swallowtail}$, V_{cusp}, and again V_{fold} rotated

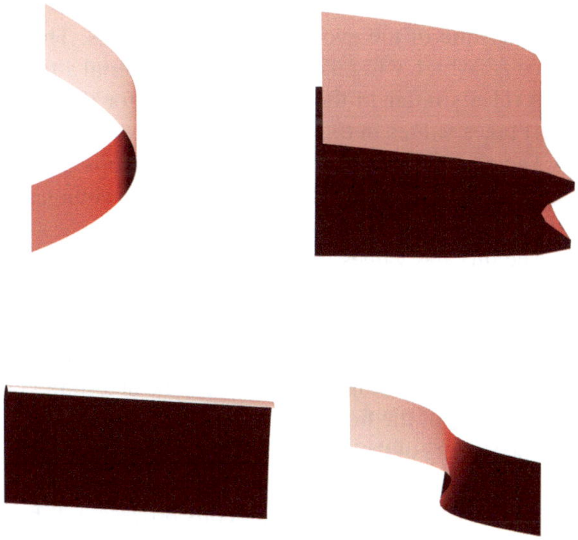

be identified with a function defined on the whole space, and supported in U. In particular, the space of probability measures it has been introduced the so-called Wasserstein distance, which allow to speak about curves or even geodesic in the space of measures on a very general space (see for example [34]).

At the same time, the identification of operators with their fundamental solution (or Gauss transform) is also very useful for the sake of visualization. In fact, every differential operator can be visualized by means of level set curves of the associated fundamental solution (or Gauss transform). In Fig. 4.22, for example, a heterogeneous assemblage of differential operators is visualized with this technique. Qualitative patterns of level lines are visualized. We have already encountered some of these patterns (see, for example, Figs. 4.5, 4.12 and 4.13) but many others are present in the image.

4.5 The Heterogenetic Flow and Its Vibrational Modes: Plateaus

In the framework constructed up to now the becoming of forms emerges as the actualisation u of the flow:

$$\partial_t u_\tau(t) = \tilde{A}_\tau(u_\tau)(t),$$

where A is the assemblage operator, t is Khronos, the time of the actualization, and τ is the time of the evolution of assemblages. The solution u will take values in a space H, which will take into account material attributes, and it is allowed to change with rules similar to the ones described for the domain. We will also assume that $A(u)$ takes values in the same set H.

The space domain (B, F) of the solution is given a posteriori with respect to the definition of operators. If the concatenation changes, the space changes accordingly, giving rise to a morphogenesis of spaces.

The flux has values in a space H of substances changing in density from point to point. This heterogenetic flow appears to be a cloud of formed substances continuously changing in form, density, composition, and velocity.

The flow itself can generate an intrinsic reference frame without the need for any external decoding. The frame can be determined as the principal or independent components of the flow.

The vectors of the reference frame will form a harmonic embedding of the process itself. If we define with &A(u) the concatenation of singular differential operators, the embedding of the heterogeneous process will be defined by all the solutions of the spectral problem:

$$\&A(u_i) = \lambda_i u_i,$$

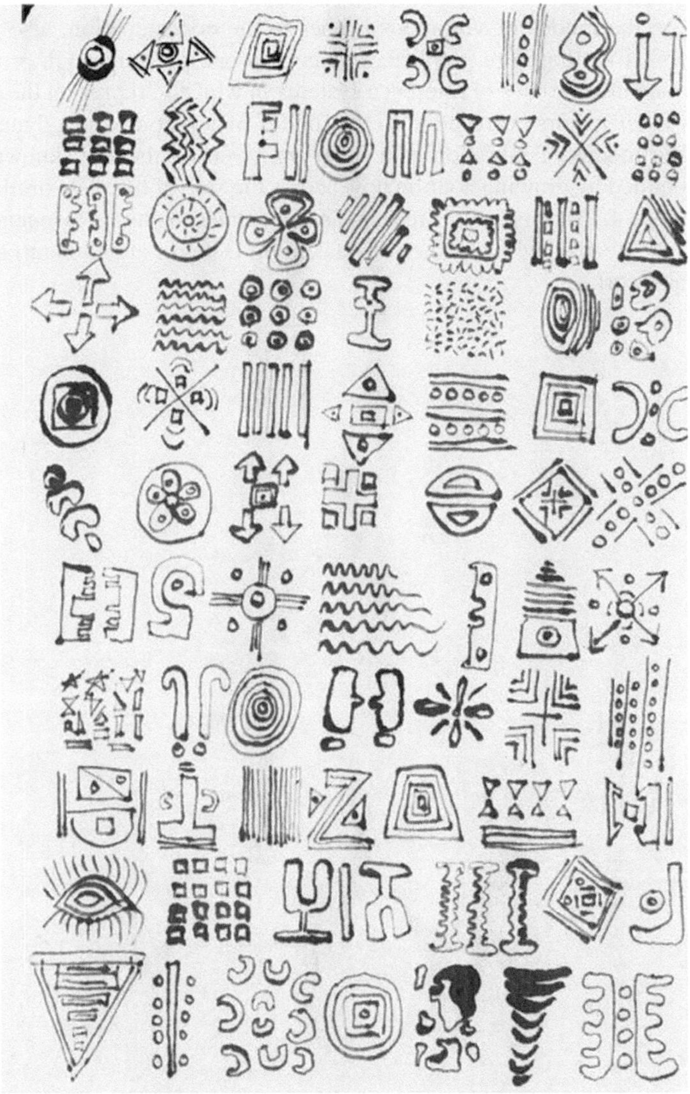

Fig. 4.22 Heterogeneous assemblage. In this figure differential operators are visualized by means of level sets of their fundamental solutions. Every operator is the generator of a different dynamics that can be recomposed on the plane of composition. Qualitative patterns of level lines are retained from the unpublished image, personal courtesy of Sergio Bianchi

where u_i are the modes of vibrations proper to the concatenation, also known as eigenvectors. It is, therefore, the heterogeneous process itself which produces (as vibrations), natural choices of reference systems in which to represent the evolution. In physics eigenvectors correspond to the modes of vibrations of a dynamics. For example, the modes of vibration of a plate are given by the well-known Chladni patterns obtained by drawing a violin bow across the side of the plate until it reaches resonance (Fig. 4.23). It is for this reason Gregory Bateson named resonance configurations as "plateaus", which is the word taken by Deleuze and Guattari as the title of their main work.

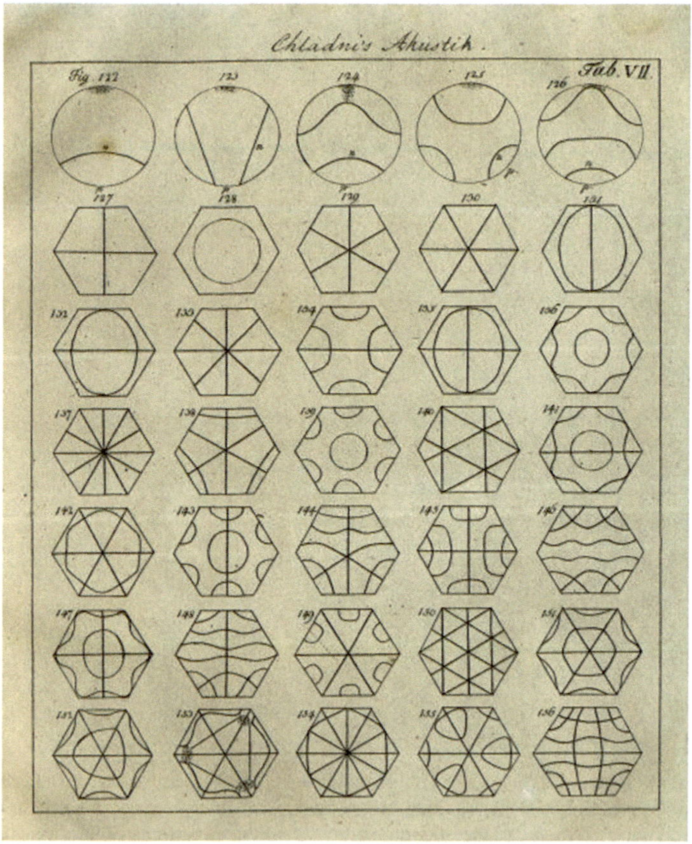

Fig. 4.23 Chladni patterns are modes of vibration of plates. Chladni's technique, first published in 1787 in his book Entdeckungen über die Theorie des Klanges ("Discoveries in the Theory of Sound"), consisted of drawing a bow over a piece of metal whose surface was lightly covered with sand. The plate was bowed until it reached resonance, when the vibration causes the sand to move and concentrate along the nodal lines where the surface is still, outlining the nodal lines. The patterns formed by these lines are what are now called Chladni figures (Wikipedia)

The instantaneous projection of the flow into its harmonic embedding is a point of the embedding and its evolution is a trajectory on the embedding.

In his book on Francis Bacon, Deleuze [8] writes that sensibility is vibration and sensation stems from the reception of these vibrations: "Sensation is vibration. We know that the egg reveals just this state of the body 'before' organic representation: axes and vectors, gradients, zones, cinematic movements and dynamic tendencies, in relation to which forms are contingent and accessory".

Commenting on this passage, Franco Berardi [2] writes:

> Like a thin film recording and decifering non-verbal impressions, sensibility allows human beings to join together ... and regress to a non-specified and non-codified state of bodies without organs that pulsate in unison." "Sensibility is the faculty of decoding intensity, which by definition means escaping the extensive dimension of verbal language. Sensibility is the ability to understand the unspoken. [2]

Interpreting literally the idea of sensitive vibration suggested by Deleuze, perception will inhabit the space of the modes of vibrations of the heterogenetic flow, i.e. its harmonic embedding. Harmonic embedding refers to the intrinsically decoding intensities of the flow without external decoding structures. This process does not correspond to a categorization but rather to a detection of the main orientations of the flow even in absence of any stabilization in fixed forms.

In [28] the authors have shown that such a harmonic approach can individuate perceptual forms from visual stimuli. Extending this approach, in Sarti and Piotrowski (2016), visual plastic formants have been individuated, showing that the principal axes determine the reference system of the space in which visual semiotic will later develop. We will reconsider these developments both in Chap. 5 devoted to cognitive neuroscience and in Chap. 6 devoted to semiotics.

We can thus emphasize that heterogeneous flow eigenvectors have a dual status: they are intrinsic reference flow axes, on one hand, and are continuously varying forms, on the other. That is they offer a pre-patterning of a possible successive stratification.

We will analyse in Chap. 6 devoted to semiogenesis the heterogenetic process in the context of an existential phenomenology. In this context the heterogenetic flow emerges as an actualization of a variety of primary "solicitations" (as developed by Merleau-Ponty 2012 (1945)), where the pre-sensory halo which "questions" and "interests" a vital principle will act as the virtual in the process of individuation. In particular, the expressive space will be individuated as the polarization along principal eigenvectors of the flow in such a way that principal modes of vibration will define the axis of body pregnancies and world saliencies, or expression and content. Now, it is precisely this dynamic of elaboration of the percepts as signifying presences that the heterogeneous approach accounts for—very precisely through the installation of harmonically correlated sub-substances in the emergence of a common empirical flow.

4.6 Multiplicity of Multiplicities

We have seen that the heterogeneous flow is able to generate modes of vibration proper to a concatenation, also known as eigenvectors. Eigenvectors of differential operators can be considered as a multiplicity of modes of coherent vibration. But eigenvectors themselves, that are functions, can become operators in a multiscale perspective.

To make a visual example: If we look at a round-dance of children, we can recognize parts of kids or single individuals as perceptual units in the scene. But we can also recognize the entire group of children as a perceptual unit in the space of previously individuated parts. To perceive the unit "group of children" it is then necessary to repeat the process of perceptual individuation at a different scale.

What this implies is that eigenvectors must be understood to have a double status: they are perceptual units defined as actualized solutions of a flow, but they are also a multiplicity of operators defining the problematic dimension of another level of individuation. In this perspective an eigenvector is at the same time the actualized solution of a differential problem and the pre-individual field of a new differential problem.

Formally, this leads us to consider a space generated by the eigenvectors. We will choose a linearly independent subfamily of eigenvectors ϕ_i and consider the generated vector space V_1. The elements of the space are now functions u_i, but they will be considered as points $\langle \phi_j, u_i \rangle$.

The space is infinite-dimensional, but in the space we can re-introduce notions of differential. We are in the context of differential calculus in Banach spaces. Due to the vector space structure, sum is allowed as well as multiplication by a scalar. In this way, if we have a functional $F : V_1 \to R$, we can define its derivative in the usual way. Let us consider a point u in the space V_1 (a function interpreted as a point). The definition of differential in this case is very much similar to the standard one.

$$F(u + th) = F(u) + dF(u)(h) + o(|h|).$$

The only difference is that the space is infinite dimensional so that we need to explicitly require that $dF(u)$ is linear and bounded, since boundedness does not follow from linearity. In other words, differential calculus in Banach space is a classical topic [5] whose goal is to extend the notion of difference in spaces of functions. Notions of manifolds and metric can be introduced, and the construction can be repeated as before for assemblages and heterogeneous flow.

Each multiplicity in this setting will be a multiplicity of multiplicities, giving rise to a modular organization as in ([6] (1980): 254). "Thus each individual is an infinite multiplicity, and the whole of Nature is a multiplicity of perfectly individuated multiplicities".

For the convenience of the reader, we have described here just one scale at a time. Indeed often, while describing a problem at a fixed scale, we are not necessarily aware that each point is indeed a multiplicity. We can consider an intuitive example from everyday life: When we plan to travel by airplane, we interpret our city of destination as a point. This point, however, is a multiplicity of streets and indeed, we will treat our destination as a multiplicity once we've arrived there and we need to reach a specific place. Again, we can think of the emergence of signs out of unstabilized and mutant perceptual experience; or likewise, of the emergence of enunciation from a multiplicity of signs. This last topic regarding the dynamics of the process of enunciation and its relationship with heterogenesis has been faced in an extended and deep way in [22]. More generally, in heterogenetic dynamics different scales are always considered, each of which is in continuous evolution. The evolution is carried in parallel and the different processes are not independent, but continuously influence one another.

References

1. Bailly, F., Longo, G.: The Physical Singularity of Life. Imperial College Press, London (2008)
2. Berardi, F.: And: Phenomenology of the End. MIT Press, London (2015)
3. Bony, J.-M.: Principe du maximum, inégalité de Harnack et unicité du problème de Cauchy pour les opérateurs elliptiques dégénérés. Annales de l'Institut Fourier **19**, 227–304 (1969)
4. Citti, G., Sarti, A.: A cortical based model of perceptual completion in the roto-translation space. J. Math. Imaging Vis. **24**(3), 307–326 (2006)
5. Deimling, K.: Ordinary Differential Equations in Banach Spaces. Lecture Notes in Mathematics. Springer, Berlin (1977)
6. Deleuze, G., Guattari, F.: A thousand plateaus. In: Capitalism and Schizophrenia, vol. 2. University of Minnesota Press, Minneapolis, 1987 (1980)
7. Deleuze, G.: Empiricism and Subjectivity. Columbia University Press, New York, 2001 (1953)
8. Deleuze, G.: The Logic of Sensation. Continuum, London, 2003 (1981)
9. Duits, R., Franken, E.M.: Left invariant parabolic evolution equations on SE(2) and contour enhancement via invertible orientation scores, part I: Linear left-invariant diffusion equations on SE(2). Q. Appl. Math. **68**, 255–292 (2010)
10. Fokker, A.D.: Ann. Phys. (Leipzig) **43**, 812 (1914)
11. Folland, G.B.: Subelliptic estimates and function spaces on nilpotent Lie groups. Ark. Mat. **13**, 161–207 (1975)
12. Guattari, F.: Chaosmosis, An Ethico-Aesthetic Paradigm. Trans. Paul Bains and Julian Pefanis, Bloomington & Indianapolis. Indiana University Press, 1995 (1992)
13. Guattari, F.: L'hétérogenèse machinique, Chimères. Revue des schizoanalyses Année **11**, 78–97 (1991)
14. Hoffman, W.C.: The visual cortex is a contact bundle. Appl. Math. Comput. **32**, 137–167 (1989)
15. Hörmander, L.: Hypoelliptic second order differential equations. Acta Math. **119**, 147–171 (1967)
16. Jedrzejewski, F.: Hétérogenese et consistance ontologique chez Deleuze et Guattari. In: Sarti, A., Longo, A., (eds.) "Differential Heterogenesis", La Deleuziana n.11 (2020)
17. Kolmogorov, A.N.: Zufällige Bewegungen. Ann. Math. 35, 116–117 (1934)
18. Longo, G., Montevil, M.: Perspective on Organisms: Biological time, Symmetries, and Singularities. Springer, Berlin (2014)
19. Lupo, D., Morawetz, C.S., Payne, K.R.: On closed boundary value problems for equations of mixed elliptic-hyperbolic type. Commun. Pure Appl. Math. **60**, 1319–1348 (2007)
20. Nagel, A., Stein, E.M., Wainger, S.: Balls and metrics defined by vector fields. I basic properties. Acta Math. **155**, 103–147 (1985)
21. Noë, A.: Action in Perception. The MIT Press, Cambridge MA (2004)

22. Paolucci, C.: A radical enactivist approach to social cognition. In: Pennisi, A., Falzone, A. (eds.), The Extended Theory of Cognitive Creativity: Interdisciplinary Approaches to Performativity, pp. 59–74. Springer (2020)
23. Petitot, J., Tondut, Y.: Vers une neurogéométrie: Fibrations corticales, structures de contact et contours subjectifs modaux. Mathématiques et Sciences humaines **145**, 5–101 (1999)
24. Picone, P.M.: Teoremi di unicità nei problemi dei valori al contorno per le equazioni ellittiche e paraboliche, Atti Accad. Naz. Lincei Cl. Sci. Fis. Mat. Natur. Rend. **22**, 275–282 (1913)
25. Planck, M.: Sitzungsber. Preuss. Akad. Wiss. Phys. Math. Kl. 324 (1917)
26. Rothschild, L., Stein, E.M.: Hypoelliptic differential operators and nilpotent groups. Acta Math. **137**, 247–320 (1976)
27. Rouvroy, A.: The End(s) of Critique: Data Behaviourism versus Due Process. In: Hildebrandt, M., de Vries, E. (eds.) Privacy, Due Process and the Computational Turn, pp. 143–167. Routledge, London (2012)
28. Sarti, A., Citti, G.: The constitution of perceptual units in the functional architecture of V1. J. Comput. Neurosci. **38**(2), 285–300 (2015)
29. Sarti, A., Longo, A. (Eds.): Differential Heterogenesis: Deleuze, Mathematics and the Creation of Forms, La Deleuziana n.11 (2020)
30. Sarti, A., Citti, G., Petitot, J.: The symplectic structure of the primary visual cortex. Biol. Cybern. **98**, 33–48 (2008)
31. Strichartz, R.S.: Sub-Riemannian geometry. J. Differ. Geom. **24**, 221–263 (1986); Correction, ibid. 30, 595–596
32. Thom, R.: Morfologia del semiotico, (Paolo Fabbri Editor), Meltemi Editore, Roma, 2006 (1981–1990)
33. Turing, A.M.: The chemical basis of morphogenesis. Philos. Trans. R. Soc. London. Ser. B, Biol. Sci.**237**(641), 37–72 (1952)
34. Villani, C.: Optimal Transport: Old and New. Springer (2008)

Chapter 5
Differential Cognitive Neuroscience

5.1 Neuromagma

We will begin this section by considering the differential geometry of the cerebral cortex from the point of view of heterogenesis. Rather than considering the cortex as a network, as is frequently done, it will be regarded as a neuromagma of differential constraints defined on an assemblage of manifolds. Every point of the manifold will eventually be joined to other points with integral curves always tangent to admissible distributions of planes generated by vector fields. The dynamics emerging from this virtuality will be closer to a heterogeneous version of field theory than to the connective dynamics of either artificial neural networks, digital networks, or actor-networks. The principal difference with respect to these models will be the presence of material uncertainty due to the non-commutativity of fields and the capability of overcoming the rigidity of coded communication while maintaining a radical heterogeneity of the virtual. In fact, due to the very anisotropic character of neural assemblies, cortical space is constituted by assemblages in sub-Riemannian geometry. In particular, the capacity of sub-Riemannian geometry to deal with non-commutative fields allows for the introduction of uncertainty principles at all scales, as well as for a variety of groups that extend well beyond Heisenberg uncertainty.

In the following, we will outline the evolution of neurogeometrical models of the cortex, starting from the first models, based on group theory, which are homogeneous in space and time, progressing through models showing differential heterogeneity in space alone, and ultimately arriving at models of embodied plasticity in which a full deployment of spatial and temporal heterogeneity is put in place, opening onto the creation of new geometries and new possibility spaces. We are especially interested in considering perception and imagination as processes of composition of assemblages, in such a manner as to include the possibility of adding, subtracting and modifying differential fields that constitute the immanent plane of composition of cortical structures. Percepts will be considered as eigenvectors or modes of vibration of differential assemblages with its own internal heterogeneity. In fact, percepts invariably emerge with different modalities—whether pertaining to

© Springer Nature Switzerland AG 2022

A. Sarti et al., *Differential Heterogenesis*, Lecture Notes in Morphogenesis,
https://doi.org/10.1007/978-3-030-97797-9_5

saliency and pregnancy, to visual and motor components—which are at the base of the cosubstantiality of semiotic phenomena. In this context, harmonic analysis will permit to group different components and diversify planes of expression and content so as to construct expressive spaces. Let's now try to enter into greater detail in our considerations of the differential structures and assemblages of the neuromagma.

5.2 Structures, Assemblages, and Plateaus of the Neuromagma

5.2.1 Neurogeometry

The primary visual cortex (V1) is a part of the brain located behind the head where a first cortical processing takes place. The visual stimulus arrives at the cortex after that it has been projected into the retinal plane, transduced into an electrical signal by the retina, propagated by the optic nerve to the lateral geniculate nucleus and finally projected into the cortex. V1 has a modular structure, which was discovered in the 80s by the Nobel Prize winners David Hubel and Torsten Wiesel. Modular means that for each point of the retinal plane, there is a complete family of cells which are sensitive to different features of the stimulus, like orientation, scale, frequency, disparity, colour, speed of movement, and many others, This set of cells constitutes the so-called hypercolumnar module.

We can model this hypercolumnar structure as a manifold (x, y, f) where x, y is the position of the cell on the retinal plane and f represents a specific feature selected by the family of cells. This space is specific for every family of cells and it is equipped with a non commutative Lie group structure whose properties depend on the invariances of the feature encoded by the cells. Consequently, the structure has been described with the instruments introduced in the previous chapter. It has been experimentally observed that each cell is connected with neighbouring cells with a very specific connectivity pattern, which differs from one family of cells to another. Propagation of the visual signal takes place along these connectivity patterns.

In a series of papers,[1] we introduced suitable sub-Riemannian structures, different for each family of cells, to model the interaction field between cells, considered as points of the manifold. The choice of the admissible tangent plane AT at every point p, allows to describe the direction of propagation; it also offers a good model of cortical connectivity. The directions of propagation of the visual signal are described

[1] Odd simple cells have been modelled in the rototranslation group [8], even simple cells in the symplectic group [35], complex cells sensitive to motion in the Galilean group [3], cells sensitive to curvature in the Engel group [1], simple cells with rotation and scale in the Heisenberg group [4], hyperbolic structures on connectivity have been studied in Sarti et al. [32], metric deformations induced by the stimulus in Franceschiello et al. [15, 16]. The interested reader could refer to [9] for a review.

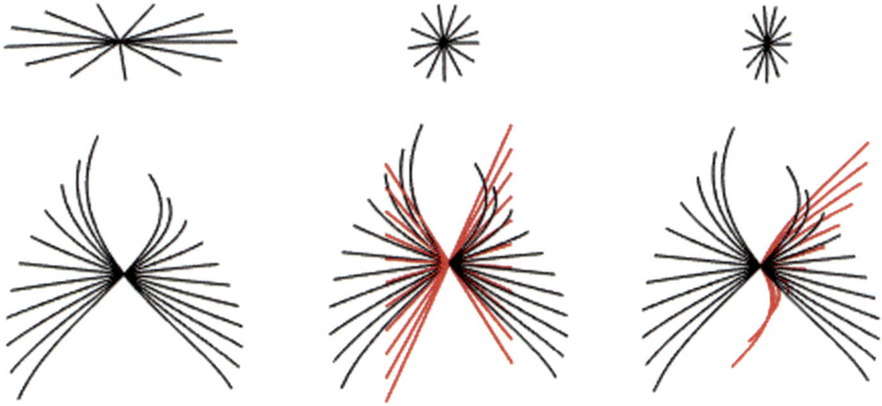

Fig. 5.1 Integral curves of fields of different families of cells. Upper: LGN cells with different elongation. Bottom: V1 odd simple cells (left), even simple cells (centre), T-junction complex cells (right). Every pattern is composed by integral curves of vector fields that are generators of a specific Lie algebra

by the entire set of integral curves of a basis of vector fields $(X_1, \ldots, X_m)$ of AT at every point.

With the same family of cells, the integral curves of fields related to each cell is the same, up to a group action (Fig. 5.2). Cells belonging to different families, and characterized by selectivity of different feature f, have different integral curves. Some patterns of integral curves are shown in Fig. 5.1.

For example, if we consider the cells of the Lateral Geniculate Nucleus (LGN), they are sensitive to the position (x, y) of the visual stimulus and constitute the group of translation that is the manifold R^2 equipped with a Riemannian metric. For every point (x, y) of the visual plane R^2, there is a cell sensitive to the position of the stimulus. Cells of this kind are connected by integral curves tangent to horizontal planes generated by

$$X_1 = \partial_x, \quad X_2 = \partial_y.$$

Different Riemannian metric are allowed at every point, generating different patterns of integral curves (see Fig. 5.1, upper row). Radially symmetric patterns correspond to Euclidean metrics, while elongated patterns correspond to suitable Riemannian metrics.

To make another example, following the model proposed by [8], the family of odd simple cells of V1 are sensitive to the position (x, y) of the stimulus and to its orientation θ in the manifold $R^2 \times S^1$. It can be described as the group of rotation and translation $SE(2)$, equipped with a sub-Riemannian metric and the following choice of the generators of AT at every points (x, y, θ):

$$X_1 = \cos(\theta)\partial_x + \sin(\theta)\partial_y \quad X_2 = \partial_\theta.$$

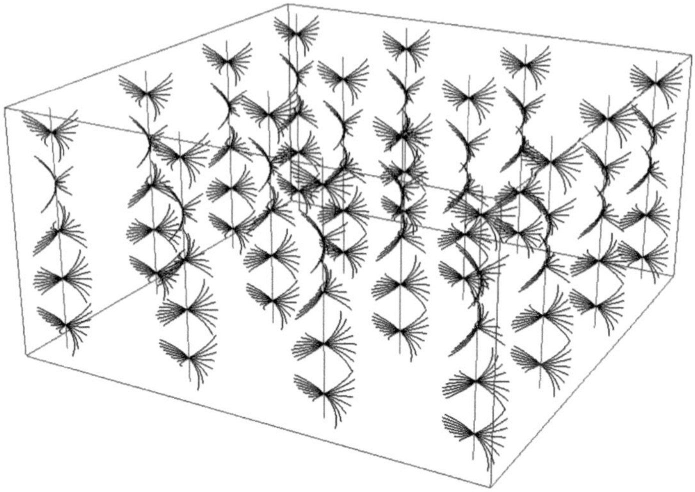

Fig. 5.2 Integral curves of fields coding the feature of position and orientation. Every pattern is composed by integral curves of vector fields generating the Lie algebra of rototranslation

Here, (x, y) is the position of the cell in the visual plane, θ is its orientation, X_1 is the orientation of the tangent to the stimuls, and X_2 corresponds to angle θ.

The structure of tangent planes is visualized in Fig. 4.8 while the corresponding integral curves of fields are shown in Fig. 5.2.

5.2.2 *Operator Kernels*

As we have seen, the wet brain is not a network, but a continuous neuromagma of many kind of cells and neurotransmitters with different dynamics. Accordingly, interaction between cells would be more appropriately expressed by continuous operators instead than connectivity curves as in the case of a network. In particular, interaction takes place as the convolution with a kernel Γ that is the fundamental solution of a certain operator A in the sub-Riemannian metric proper of a specific kind of cell:

$$\delta = A_{SR}\Gamma.$$

For example, the kernel joining odd simple cells could be given by the fundamental solution of the Fokker-Planck operator in the sub-Riemannian metric of the rototranslation group:

$$\delta = FP_{RT}\Gamma,$$

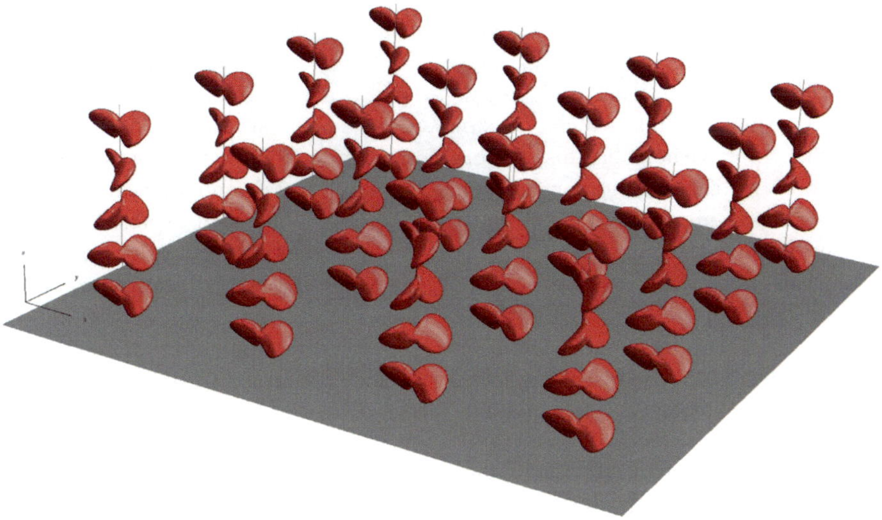

Fig. 5.3 Fokker-Planck kernels of the group of rototranslation

that is

$$\delta = (X_1 + \sigma^2 X_{22})\Gamma.$$

These kernels (Fig. 5.3) are like fat versions of the integral curves of fields as visualized in Fig. 5.2.

Or, for complex cells sensitive to curvature, the conjunction kernel could be given by the fundamental solution of the Fokker-Planck operator in the sub-Riemannian metric of the Engel group:

$$\delta = \Delta_E \Gamma,$$

that is

$$\delta = (X_{11} + X_{44})\Gamma.$$

Of course, every operator and geometry induces a specific kind of kernel acting in a specific symmetry group.

In case of interaction between heterogeneous cells, the sub-Riemannian structure is the one of the neurogeometric assemblage we have considered in the previous chapter, leading for example, to Fig. 4.18.

$$\delta = A_{SRa}\Gamma$$

Notice that the kernels depends both on the sub-Riemannian geometry SR and on the operator dynamics A and that the assemblage is a local composition, maintaining the heterogeneity of the space.

Fig. 5.4 Multiple kernel assemblage

In the neuromagmatic structure of the brain multiple kernel assemblage take place, since the same family of cells interacts with different families of cells, giving rise to different assemblages in each case (see Fig. 5.4).

5.2.3 *Brain Activity and Plateaus*

If a certain cell is solicited by a stimulus, then it enters in a preactivation state. In this state, the activity of the cell depends on the activity of others cells to which it is connected. If Γ is the operator kernel of the cell, the asymptotic activity a of the cell is given by

$$a(x) = \Gamma(x, x') * a(x').$$

Since the stimulus contains all kind of features in a very singular composition it preactivates a constellation of cells sensitive to different features. Or in other words, it selects a very specific constellation of operators kernels at points x_i (in red in Fig. 5.5).

The overall set of kernels is contained in the adjacency matrix M_{ij}, that is the assemblage of all of them: $M_{ij} = \Gamma(x_i, x_j)$.

This set of preactivated cells begins to communicate with each other with its own dynamics, giving rise to an emerging global dynamic. In Sarti and Citti (2014), it is shown that eigenvectors of the assemblage matrix M correspond to organized units, i.e. to the most coherent form in the stimulus image:

$$M\tilde{a} = \lambda_k \tilde{a}.$$

In physics, eigenvectors correspond to the modes of vibrations of a dynamics. For example, the modes of vibration of a plate are given by the well known Chladni patterns obtained by drawing a violin bow across the side of the plate until it reaches

Fig. 5.5 The constellation of operator kernels selected by a stimulus are represented through their integral lines (in red)

resonance. For this reason, Gregory Bateson named resonance configurations as "plateaus"—a word that of course was taken by Deleuze and Guattari in the title of their main work. Analogously, perceptual units are resonance configurations of the operators assemblage excited by the stimulus. We therefore have an individuation of forms by harmonic analysis of the assemblage matrix M_{ij}. The individuation of forms is made possible on the base of spectral clustering and dimensionality reduction. In [30], the problem of perceptual grouping has been faced in terms of reduction of the complexity in the description of a scene. The visual scene is described in terms of the matrix M_{ij} with a complexity of order $O(N^2)$ if N discrete elements are present in the scene. The idea of Perona and Freeman is to describe the scene approximating the matrix by the sum of matrices of rank 1 and complexity N, each of which will identify a perceptual unit in the scene. If the number of the perceptual units present in the scene is much smaller than N, this procedure reduces the dimensionality of the description. A rank 1 matrix will be represented as the external product of a vector p with itself. The first one will be computed as the best approximation of M_{ij} minimizing the Frobenius norm as follows:

$$p_1 = argmin_{\hat{p}} \sum_{i,j=1}^{N} (M_{ij} - \hat{p}_i \hat{p}_j)^2,$$

where the term $pp^T = \sum_{i,j=1}^{N} \hat{p}_i \hat{p}_j$ is the rank 1 matrix with complexity order $O(N)$. Perona proved that the minimizer p_1 is the first eigenvector v_1 of the matrix M with largest eigenvalue $\lambda_1 : p_1 = \lambda_1^{1/2} v_1$. Accordingly, the problem is repeated on the vector space orthogonal to p1. The minimizer will correspond to the second eigenvector, and iteratively the others eigenvectors are recovered. The process ends when the associated eigenvalue is sufficiently small. In this way, in general, only n eigenvectors are selected, with n < N, leading to the dimensionality reduction. The problem of grouping is thus reduced to the spectral analysis of the affinity matrixM_{ij}, where the salient objects in the scene correspond to the eigenvectors with largest eigenvalues. We've demonstrated in the previous paragraphs that this spectral analysis can be implemented by the neural population in the functional architecture of the primary visual cortex. We can now interpret eigenvectors of the assemblage as the perceptual units segmenting the scene.

5.2.4 Plateaus I: The Formemes of Plastic Forms

In *Traité du signe visuel* [20] the Groupe μ analyses the basic forms that are perceived as when only a background is present in the visual scene. The status of the background is very peculiar, since it is not considered as undifferentiated and unlimited or without defined boundaries. The group μ defines the background as a limited space, like a sheet of paper or a painting's frame. In this manner, the background contains its own form and for this reason is defined as "paradoxical":

> *The form of the background intervenes by imposing its laws on the other forms that detach from it* [20].

The formemes indicate the relationship between figures and the paradoxical background. Particularly the first formeme corresponds to the position of the figure in the background:

> *It is clear that different significations emerge when the form is positioned in the center of a background or in a different point of the space.* [20]

The structure of tensions of the formeme of position is tripartite and it is articulated as in the following (Fig. 5.6):

- First tension: the figure could be central or marginal in relation to the centre (centrality/perifery).
- Second tension: if marginal, the figure could be displaced vertically (verticality).
- Third tension: if marginal, the figure could be displaced horizontally (laterality) The structure is represented in Fig. 3.1, from [20].

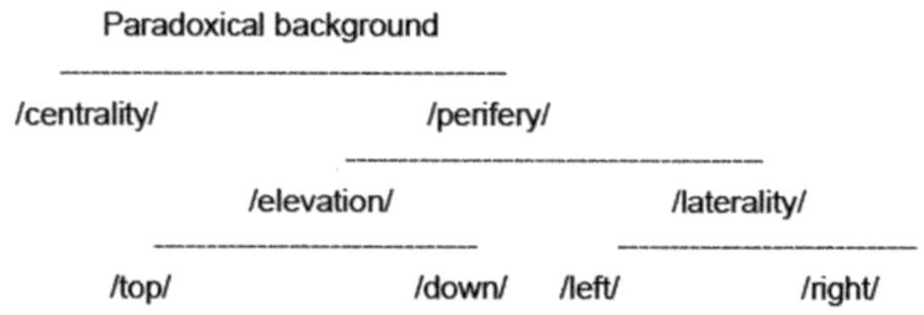

Fig. 5.6 Structure of tensions of the paradoxical background as analysed by the Group μ in [20]

Fig. 5.7 An isotropic and delimited relational field defines the paradoxical background. A few kernels (integral curves of fields) are sampled and visualized in figure

The question of emergence of formemes as actualization of a field of differential operators has been faced in [34]. In order to retrieve the basic forms of plastic semiotics let's introduce a suitable relational field for the paradoxical background, that is the virtual for the emergence of forms. The metric structure of the background is isotropic and delimited in space domain Ω (Fig. 5.7), in such a way that the conjunctive kernel Γ reads:

$$\delta = \Delta_E \Gamma,$$

where Δ_E is the Euclidean Laplacian.

By spectral decomposition of the operator assemblage in Ω, we obtain as principal eigenvectors the three basic forms of visual semiotics: centrality/perifery, verticality (top/down), and laterality (left/right). The three functions are visualized in Fig. 5.8, well representing the structure of tensions of the paradoxical background as proposed by the Group μ.

Let's notice that the three plateaus are not partitioning the space but they are *simultaneously* solutions to the harmonic problem. They constitute the modes of vibration on which eventually apparatus? of the sign can intervene to cut the space in oppositive regions. For example, a cut could separate the first formeme in a central and a peripheric region. The cutting will be the result of the morphodynamics of the sign. We will face the question of semiogenesis later in chapter, but for now, we will maintain the plateaus as a smooth tension field on a proto-expressive space.

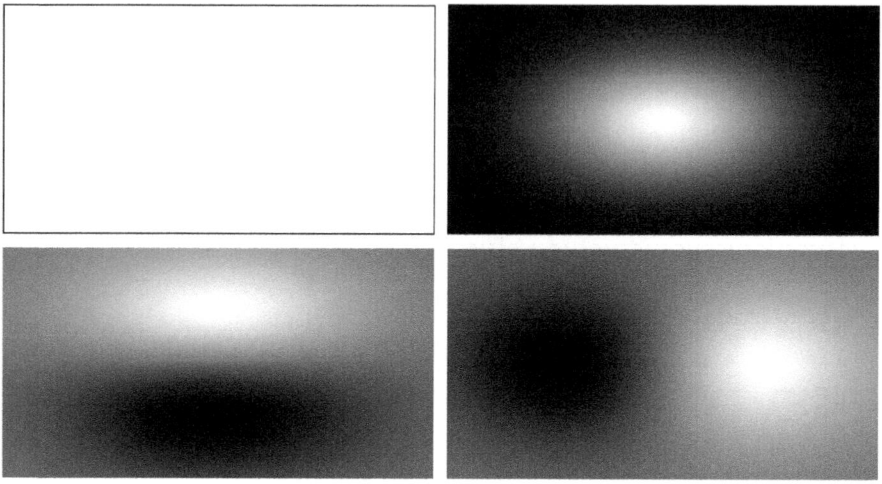

Fig. 5.8 The paradoxical background (top left) and the three main eigenvectors of the assemblage operator are shown (top right and bottom). The eigenvectors are obtained by numerical simulation. They corresponds to the three basic forms of visual semiotics: centrality/perifery, verticality (top/down), and laterality (left/right)

5.2.5 Plateaus II: Perceptual Grouping

Let's now consider a problem of grouping in which a perceptual unit emerges from a set of fragmented features. Let's consider first the case in which the basic features are homogenous as in Fig. 5.9, where a set of position-orientation elements are present.

Every element is defined in terms of its position and orientation $\xi_i = (i, i, \theta_i)$. A straight line and a curvilinear shape—within what otherwise appears as a collection of randomly placed elements. For position-orientation elements the conjunctive operator $\Gamma *$ is defined in the rototranslation group and writes:

$$\delta = \Delta_{RT} \Gamma,$$

where $\Delta_{RT} = (X_{11} + X_{22})$ is the sub-Riemannian Laplacian.

The assemblage matrix is then

$$M_{ij} = \Gamma(\xi_i, \xi_j),$$

where ξ_i and ξ_j is any couple of points of the visual stimulus conjoined by the kernel Γ. A schemata of the field $\Gamma(\xi_i, \xi_j)$ selected by the visual input is visualized in Fig. 5.10.

Spectral decomposition of the assemblage matrix yields two salient principal components corresponding to the two perceptual units. Results are visualized in figure.

Fig. 5.9 A stimulus as a random distribution of elements of position and orientation. Two figures are perceived as groupings of coherent position-orientation patterns

Fig. 5.10 Principal eigenvectors of the assemblage matrix correspond to perceptual units

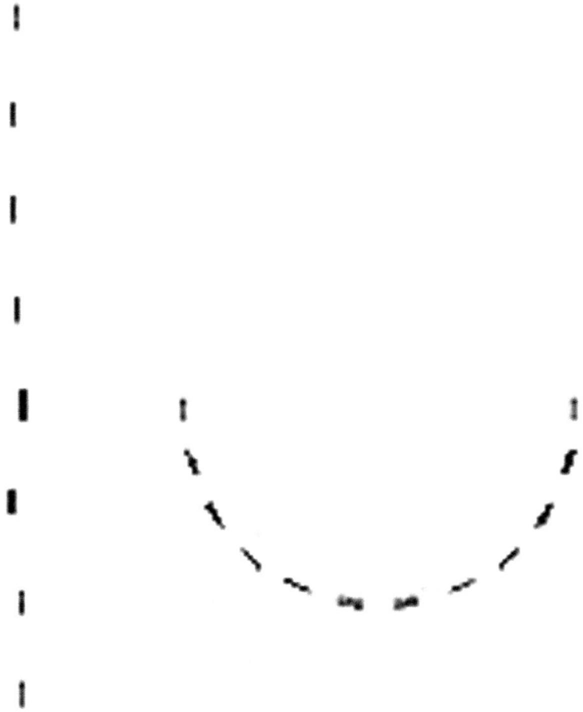

Notice that only the principal components of harmonic modes become perceptual units, while a myriad of other elements are not integrated and therefore]remain at the level of a tensive background that maintains a character of virtuality resisting any attempt at integration. This situation clearly reflects one of main points of the Simondonian conception of individuation—namely, that where integration takes place, it is always in a partial way, with an irreducible set of preindividual elements continuing to exist in a virtual modality of being. This set is disentangled from the main visual configurations or Gestalts and constitutes an active background of virtuality with the power to constitute future forms. In other words, although Gestalts are the main harmonic forms, they lose power by effect of actualization while unintegrated elements maintain the morphogenetic force that allows them to build alternative configurations. Far from mere side effects of the integration process, the dissonant elements constitute the tensive counterpart of the creation of forms and comprise a fundamental pole in the dialectics of forces and forms in the perceptual field.

5.2.6 Plateaus III: Hallucinations

Finally, if the complete neurogeometry of the visual cortex is activated, for example, by the action of psychedelic drugs (psilocybin, peyote, LSD), the eigenvectors of the overall operator (Fig. 5.11) correspond to the images of hallucinations reported in the literature, after the experiments on entoptic vision of the 60s. We see that the mechanism of formation of perceived objects is the same as the mechanism of production of hallucinations. So, the perception of shapes in our daily vision is a kind of hallucination controlled by the stimulus. Jan Koenderink, a great researcher of visual perception, had this intuition 40 years ago and today we find a confirmation of this idea.

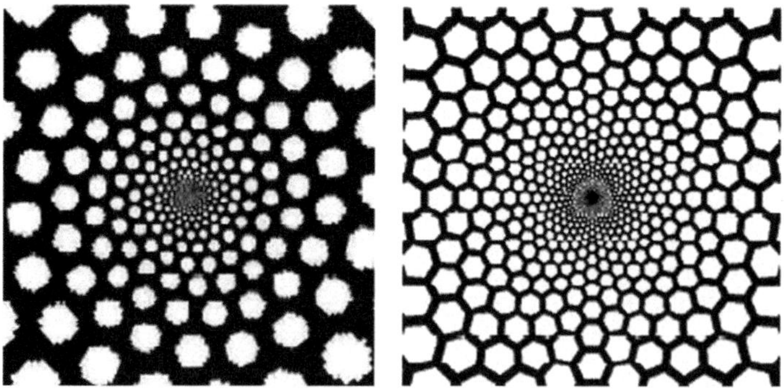

Fig. 5.11 Entoptic vision. Visual hallucinations are the harmonic modes of the neurogeometric structure of the primary visual cortex when the entire set of operator kernels is fully preactivated by using psychedelic drugs like psylocibine, peyote or LSD

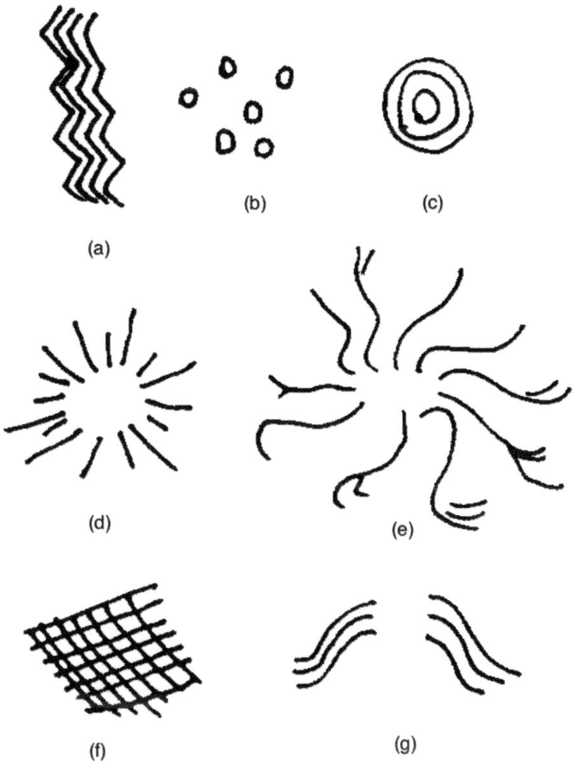

Fig. 5.12 Categories of entoptic phenomena visualized following descriptions of (H.Kluver, 1926): **a** Parallel figures, **b** Small circular figures, **c** Large circular figures, **d, e** Radiations, **f** Grids, **g** Bilateral duplication

The morphologies of hallucination are very different but many of them are quite recurrent. These have been classified be Heinrich Klüver in 1926 [24] by using mescalin and peyote as visualized in Fig. 5.12. Following his classification, there are forms with a boundary, without boundary, spider webs, spirals and many other forms that we find as harmonic principal components of the cortical neurogeometry.

These kinds of entoptic forms are very commonly found in cave art in all the continents. Spider webs and twisted snakes in particular are very common. And in fact, it was Aby Warburg, the famous art historian of the nineteenth century, who studied the psychedelic rituals of Native American of the clan of the Pueblo (North America) and particularly the ritual of the snake.

His paper "A Lecture on Serpent Ritual" [40] is the result of a lecture he gave in 1923 as a farewell address to doctors and patients at the end of one of the various stays in the clinic to which he was confined due to nervous breakdowns. This lecture, which was then published in the Warburg Institute's "Journal" in 1939, starts from a trip to the Pueblo Indians and, by emphasizing the psychic power of images, comes to grasp the main characteristics of paganism and magic. The lucid analysis of the snake ritual among the Pueblos leads Warburg to understand the symbolic analogy between the lightning bolt and the reptile. Analogy on which this tribe bases the invocation of

Fig. 5.13 From entoptic forms to archetypal morphologies: The ritual of the snake studied by Aby Warburg in the clan of the Pueblo. In figure Cleo Jurino, Cosmological drawing, from [40]

the storm through the dance with live snakes and that will be outlined also by Kluver in his analysis of entoptic forms. Warburg also makes an excursus on the presence and importance of the snake in other cultures, highlighting how this animal's power, which was experienced as an enigmatic demon, was central to various ancient cults (Fig. 5.13). One need only think of Dionysus and Asclepius in Greece, Tiamat in Babylon, the serpent of the old testament, and so on. Warburg finally shows how entoptic forms became archetypal forms in culture and in the history of art, in such a way as to link morphological fields with the symbolic forms through which man relates to himself, to nature and to the sacred.

5.3 Strata: Modal and Amodal Completion

Cortical assemblages can be organized in strata, following the modular structure of cortices. For example, the visual system is articulated in layers starting from the lateral geniculate nucleus (LGN), towards the primary visual cortex V1, until higher level cortices V2, V3, ..., V6. In the problem of completion of the Kanizsa triangle mainly the primary visual cortex and LGN are involved.

Looking at Fig. 5.14, we see the presence of a triangle with very sharp boundaries even in completely homogenous part of the image. These boundaries are called subjective by Gaetano Kanizsa [23], meaning that they are not present as gradients of the image but they are built by the visual system of the perceiving subject. If the inducers of the triangle are not aligned, as in the present case, boundaries don't appear straight but curvilinear. Moreover, the triangle surface is perceived as brighter than the background. The triangle is then modally completed, meaning completed with the modality of vision (since we see phenomenally the brightness of its surface), while the three disks are a-modally completed, meaning that we know that they are partially occluded by the triangle but we don't see the occluded surface. Two main processes are thus at stake in the perception of the scene: the propagation of the boundaries of the triangle and the filling in of its surface. The visual system implements these processes in such a way that V1 is engaged to accomplish the completion of missing boundaries while LGN performs the filling in of the interior of figures. The relation between the two processes is the one between particle and field in Gauge field theory, as outlined in [10].

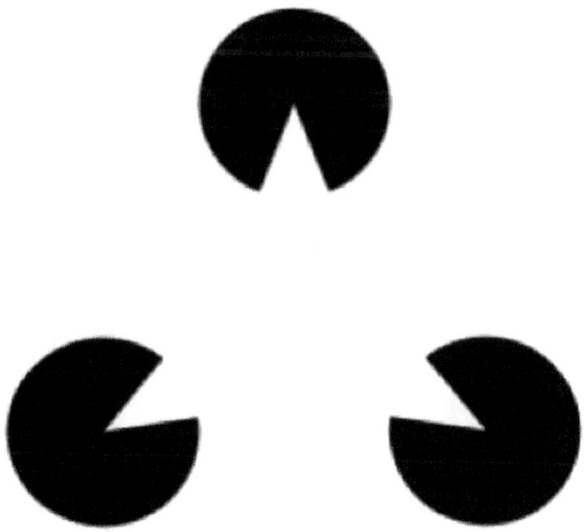

Fig. 5.14 The Kanizsa triangle: the formation of curvilinear subjective boundaries as well as of the subjective surface brighter than the background is due to the conjoint action of the two strata of LGN and V1

It is important to note that boundary completion is implemented in the 3D sub-Riemannian geometry of the rototranslation group, proper of the functional architecture of V1 simple cells, while filling in of the surface image and contrast propagates in the 2D plane (in our hypothesis the LGN plane). In order to study the joint evolution of the two different geometries, we will project both geometries onto a 2D visual plane.

We will start by propagating boundaries, represented as a 2D vector field $\vec{A}$ in the 2D projection of the sub-Riemannian geometry. In order to do so, we first represent $\vec{A}$ in polar coordinates as

$$\vec{A}(x, y) = (A_1(x, y), A_2(x, y)) = |A(x, y)|(\cos(\theta(x, y), \sin(\theta(x, y))).$$

Then we project the 3D vector fields X_1 and X_2, along the vector $\vec{A}$. The vector X_2 will be discarded in the projection, while the vector field X_1 will be evaluated only at the at the orientation $\theta(x, y)$ associated to $\vec{A}$:

$$X_{1,A} = \cos(\theta(x, y))\partial_x + \sin(\theta(x, y))\partial_y.$$

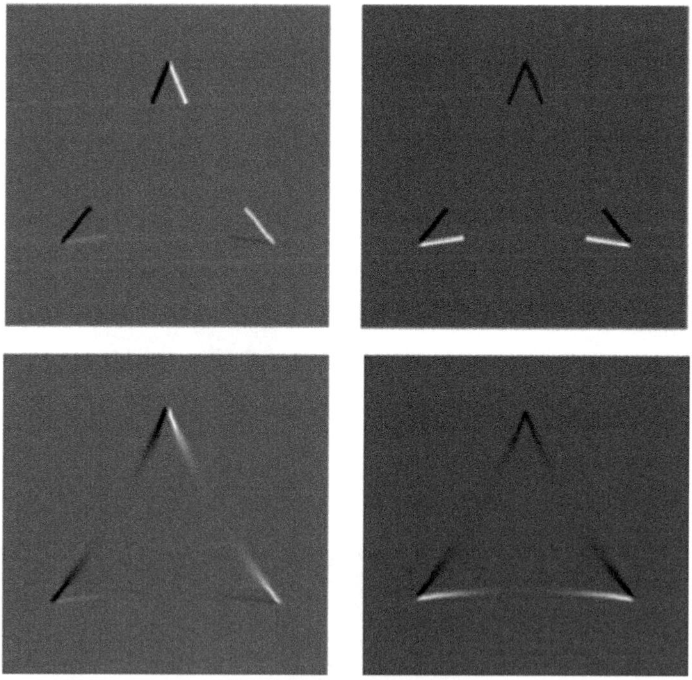

Fig. 5.15 First stratum: Emergence of subjective boundaries as propagation of the gradient of triangle inducers (top, respectively, x and y components) in the tangent direction to complete boundaries (bottom, respectively, x and y components)

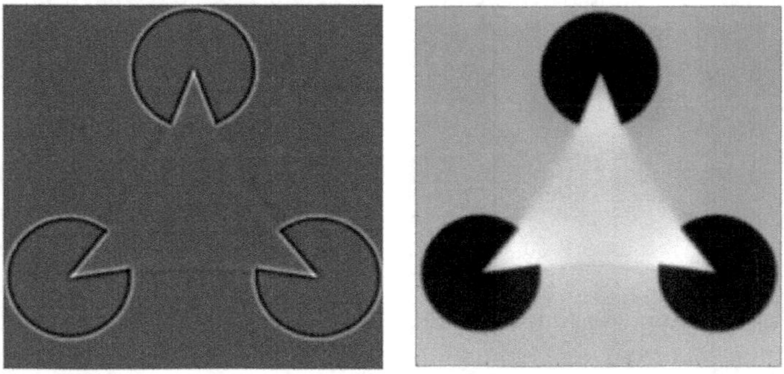

Fig. 5.16 Second stratum: Filling in of subjective boundaries (left) to accomplish Kanizsa subjective surfaces (right)

The corresponding propagation will become

$$X^2_{11,A}\vec{A} = -\nabla\phi,$$

where the vector field $\vec{A}$ representing the boundaries is propagated in the direction of the tangent vector X_{1A} and forced by the inducers of the triangle $\nabla\phi$. The A_1 and A_2 components of the vector fields are visualized in Fig. 5.15. See [9] for more details.

At the same time, filling in from boundaries to surfaces is performed by the 2D functional geometry of LGN and the final perceived image ϕ is given by

$$\Delta\phi = \frac{1}{2}(\Delta h + div(\vec{A})),$$

denoting the propagation of the boundaries of the initial image Δh and of the subjective boundaries $div(\vec{A})$ to create subjective surfaces. The result of filling in is visualized in Fig. 5.16.

5.4 Composition-Actualization of Percepts

Cortical cells can be represented by derivations so that their action will be the one to derive the visual stimulus point by point as in [25]. Looking at the cortical surface, in every point p there is a different operator of the type:

$$A(u)(p) = \sum_i X_i^{\beta(p)}(n_i(p)X_i^{\beta(p)}u)(p),$$

where X_i is an arbitrary directional derivative, $n_i(p)$ is a weight on X_i and $\beta(p)$ is an arbitrary order of differentiation. In this way, we have an operator with different

Fig. 5.17 Spatial heterogeneity of differential operators. Left: Random distribution of operators on the cortical surface. The Gauss transform of operators is visualized. Right: Differential operators selected by the image out of a fiber of operators

orientation (eventually isotropic) and different order at every point. In some species $n_i(p)$ is a smooth map, giving rise to the so-called pinwheel structure, in other species it is just a random distribution.

Let us consider here a totally heterogeneous distribution as a throw of the dice of operators. The operators are visualized in Fig. 5.17 via their Gaussian transform. Hence, they are represented as a family of kernels.

The output of this distribution of cells in response to the visual input I will be

$$A(I)(p) = \sum_i X_i^{\beta(p)}(n_i(p)X_i^{\beta(p)}I)(p).$$

Fig. 5.18 Emergence of percepts as actualization of heterogeneous differential operators in the cortical surface: **a** Differentiation

We see that the stimulus image is completely destroyed by the derivations (Fig. 5.18) and it is still an open problem in neuroscience to understand if the perceived image can be reconstructed. We will propose here a heterogenetic procedure to construct the perceived image.

In fact, we can consider the flow

$$\partial_t u(p) = A(u)(p) - A(I)(p),$$

where the operator $A(u)(p)$ is actualized by means of the evolution in the Khronos time t.

Fig. 5.19 Emergence of percepts as actualization of heterogeneous differential operators in the cortical surface: **b** Various stages of integration

Written in this form it can be identified with the first variation of a functional

$$F(u) = \sum_i \int (n_i(p,t)(X_i^{\beta(p)} u - X_i^{\beta(p)} I)(p))^2,$$

with random coefficients $n_i(p)$.

As shown in Fig. 5.19, this process constructs the perceived image, that is the original stimulus I up to an Harmonic function, as in the previous section, where harmonic means no more the annulation of the Laplacian but the annulation of the heterogenous operator.

5.4.1 The Four Stages of the Constitution of Plastic Morphologies

Visual perception involves all the strategies we have presented up to now and many others. All the complexity of perception appears, for example, in front of a painting like the pregnant Madonna of Piero della Francesca. Here the extent of differential heterogeneity is enormous even if we focus only on the first layer of the perceptual process, which Algirda Greimas refers to as the semiotics of visual plastic [19]. The tensions induced by the background and by the internal gradients of the image, as well as the construction of coherent boundaries and surfaces of figures, pop up in our consciousness as perceptual units that actualize the virtual assemblage of preactivated cells. The pregnant Madonna, the accompanying angels, the embroidered tents, the pleated clothes, the shaded hands and all related tension fields emerge then as plateaus of the virtual assemblage. Modal and a-modal completion effects are always present to detect and complete occluded objects. The virtual assemblage will then contain concatenations of Riemannian and sub-Riemannian patches, in addition to various other visual features. Some samples of heterogeneous kernels involved in the virtual assemblage are schematized in Fig. 5.20. Of course, the entire perceptual process involves a much richer stratification of the virtual, which implicates embodied, enactive and socio-cultural dimensions.

But already at this early level of the semiotics of the visual plastic we encounter the four stages of the constitution of perceptual forms:

- First the stimulus is a dust of information which is made by a set of microelements, for example photons or pixels.
- This dust preactivates a constellation of cells meaning that they are set in a state where they can deploy their own dynamics.
- The preactivated cells are put in contact with one another by neural connectivity. In this way, a constellation of differential operators is composed which constitute a heterogeneous differential assemblage.
- The operator assemblage is then actualized giving rise to the perceived image or to salient morphological configurations. In the last case, solutions are defined by

Fig. 5.20 The pregnant Madonna of Piero della Francesca: Fields integral curves of the visual plastic are shown in figure. Just a few samples of them are visualized out of a dense magma. The phase space is too large and structured to be visualized in figure

harmonic analysis of the operator assemblage, which means considering the eigenvectors of the assemblage that correspond to the emerging state of global neural activity. The emergence of plastic forms in our perceptual system is correlated to a synchronization of the dynamic activity of the brain, corresponding to a sort of overall "vibration" of the differential assemblage selected by the visual stimulus. Specifically, the mode of vibration, or plateaus, of the assemblage correspond to the most salient morphological configurations.

Any character of cohesion of a form, including the presence of internal tensions and of field effects, which are at the centre of all phenomenology of perception, are due to the fact that each perceived morphology is the solution of a differential problem, i.e. it is a global solution obtained by integration of an assemblage of intensive elements. More explicitly, we see here *the differential origin of any morphology,* starting from the seminal idea of Goethe to the Berliner Gestalt school and subsequent evolution of Husserl and Merleau-Ponty's phenomenology of perception. We see also that the concatenation body/brain offers the set of differential constraints that can be selected by the stimulus to build assemblages, meaning that the body/brain work as the plane of composition for the emergence of any perception. We will see in the following that the set of local differential constraints are far from given a priori, as they are in fact produced and plastically moulded by subjective experience. To this point, we have only considered the salient component of perceptual experience; however, as any experience is embodied, the presence of bodily pregnancies will allow us to understand the full deployment of perception as a process that is originarily semiotic/that is semiotic in its very essence.

5.5 Embodied, Embedded, Enactive, Extended Cognition

5.5.1 Embodied Plasticity: Saliences and Pregnancies

Neuroplasticity is the fundamental characteristic of the brain. It consists in the ability of the brain to change continuously throughout an individual's life. Given that we consider neural structures to be assemblages of differential operators, considering plasticity means taking into account how an assemblage is continuously reshaped in time; or in other words, considering the perpetual metamorphosis of the virtual.

To make the things simple, let's return to the primary visual cortex. If we observe carefully simple cells in the visual cortex, we find that they are all different. For every point of the retinal plane, there is an ensemble of cells which always present different shapes and different patterns of connectivity. The hypothesis of group symmetry posed by Lie group-based neurogeometry is thus only weakly fulfilled, as the measured connectivity patterns are deformed with respect to the symmetric case. If we repeat the same measurement of the shape after a certain time, furthermore, we find that the shape is changed and that it depends on the history of stimuli that the subject has perceived.

For example, if the subject perceived mostly vertical figures, the connectivity pattern of cells will be mostly vertically oriented. This simple experiment shows that simple cells change shapes due to the plasticity of cortical connectivity, meaning that the virtual plane engendered by the brain changes geometry.

But plasticity does not depend simply on the morphology of the stimulus but also on the meaning that it produces. In fact, we can experimentally observe that the changing of neurogeometry is much more present when the stimulus has a meaning for the subject. That is, if the stimulus engages the affective, sexual, behavioural sphere of the subject, then the plastic variation is more evident. Alternatively, if there is no interest in the stimulus, little or no variation will be observed.

This phenomenon can be clearly observed in the experiments of Norman Weinberger [41], who shows that the sensitivity of cells to the frequency of stimulus changes completely if the subject is submitted to conditional learning, that is, when the stimuli are reinforced by reward or punishment. The experiments of Weinberger show that after learning the sensitivity of cells become maximal for the frequency belonging to significant stimuli. Strong deformations of connectivity kernels are introduced by the presence of pregnancies (Fig. 5.21). In this way, the morphology of cells becomes a memory of significant stimuli and the cells themselves become ab initio forms of value [33].

The process of constitution of forms of values has been described very precisely by the mathematician René Thom in terms of emergence of saliences and pregnancies. Particularly in [36], the author describes saliency in these terms (our tr.):

> Some sensory stimuli impose themselves because of their unexpected and discontinuous character. They are figures that emerge from an undifferentiated background, like the tinkling of a bell or a flash of light. I will call these sensory events that affect the senses, salient, and saliency their violent and suddenly discontinuous character. (pag. 93)

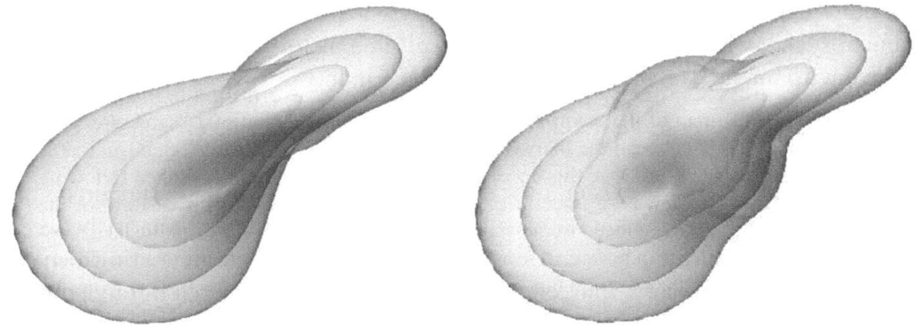

Fig. 5.21 Kernels deformed by pregnancies. Left: Kernel without pregnancies shows group symmetries. Right: Kernel after reinforcement by pregnancies are deformed like Barocco pearls. Deformation is largely unpredictable and escapes neurogeometry in symmetry groups. The final shape depends on the meaning of the visual experience in its history

Saliency, in this case, is a form which becomes relevant due to its morphological consistency, which corresponds to a kind of "gestalt" in Kohler and Wertheimer's theories. This definition applies not only to visual forms, but also tactile, auditory and general sense forms. Pregnants, on the other hand, are the forms filled with intense biological value—hunger, fear, sexual desire, etc.—imbued with profound and lasting emotional impacts of attraction and repulsion:

> ...some perceived forms, particularly because they are biologically significant, will exert effects (attraction or repulsion) on the subject expected to be long-term, such as the forms of prey, predators, sexual partners, etc. I will call pregnant these meaningful forms, and pregnancy the character associated with them. The pregnant ones, in short, are forms that give rise to a specific motor behavior, associated with major hormonal changes in the metabolism. (ibid.: 56)

In this schema, pregnancy invests saliency in the same manner as a signified invests its signifier. For Thom, this mechanism is the basis of *symbolic function in the animal psyche*. However, it must be outlined here that this conception stays within an interpretative framework, where meaning and symbol are built each on its own side and later associated. In opposition to such a logic, we will develop in what follows what we refer to as the expressive problematic, according to which signifier and the signified do not exist independently of one another and cannot even be separated in the mind.

As Thom himself outlines, we can already see this phenomenon in Pavlov's behaviourist experiments:

> It is therefore legitimate to argue that, in the situation created by Pavlov's experiment, the tinkle of the bell, for the dog, is a sign of meat. It is the primitive form of symbolic reference, which I will represent by using the notation A − > B. The formula indicates that A is a sign (stands for) for B. Similarly I shall designate with R(A) the set of psycho-physiological reactions aroused in the subject by the perception of the form A. So if we have A − > B, A sign (stands for) for B, then it is reasonable to argue that R(A), R (B), or rather that the first set is included in the second as a subset (given that pregnancy has less effects in A than in B). (ibid.: 94)

Reinforced learning would therefore be the basis of the symbolic function in the animal psyche, which primarily manifests itself in a distortion of the spatial coding of the environment. The investment of pregnancy induces

> a distortion of the map of the environment that leads to the creation of a "symbolic turn" ("anse symbolique") between the body of the subject and the body of the organism that is the source of pregnancy ... Once this encounter has happened, the turn is broken following a "disappearance cycle" whose trace in the subject's body is only the visual form of the encountered object, now reduced to the state of memory. But as soon as significance reappears, under the influence of hormonal factors, the symbolic loop can reconstitute itself every time the subject bumps into a form that even approximately resembles the form responsible for the imprinting. This symbolic identification is simply the geometrization of the subject's desire. (ibid.: 96)

Yet the extent of reinforced learning goes far beyond the scope of Pavlovian behaviorism (in which the reflex is a biological automatism), and pushes toward proto-semiotic forms made in the very first hours of the life of an infant. Patrizia Violi in [39] reconstructs the first moments of the relationship between a mother and an infant, emphasizing the passage in which the infant's behaviour ceases to be governed by automatic biological reflexes and begins to be reinforced by sociocultural factors. The newborn's first behaviour is imitative, which is innate and cognitively explained through the action of mirror neurons [31].

This imitative behaviour provokes a reaction in the parents. If their response is positive, this reinforces successive repetitions. If instead their feedback is negative, this weakens the given behaviour. The parent's affectivity influences the baby's beliefs, becoming a modulating factor in its behaviour through reward and/or punishment. If the earliest form of imitation is innate, the repetition of imitative behaviour is mediated by adult intersubjective reinforcement, and therefore is already semiotic following Eco's definition [14], which is recalled by [39]:

> One could object to talk of semiosis and semiotic behaviour at such an early stage. However, if—following Eco (1979)—one takes semiosis as any response to the environment that is not causal and constrained by a stimulus response pattern, implying a possibility of freedom and variation, one can easily recognize the beginnings of a semiotic life. Eco called this space of freedom Space C. However minimal it might be, a Space C's existence between stimulus and response testifies to the non-deterministic character of the response and, therefore, the presence of semiotic mediation. It is precisely a behaviour's mediated nature that allows us to define it as semiotic: i.e., not reducible to causal response to a stimulus."

The naturalness of the innate behaviour is modulated by intersubjective reinforcement, constituting an initial interface between nature and cultural values. Yet again, Patrizia Violi stresses that in this situation of reinforced learning, the adult takes on the role of the interpretant of the infant's behaviour:

> If a neonate's first imitation may well be no more than 'natural', repetition of imitative behaviour, due to intersubjective reinforcement from adults, is already semiotic, because it is mediated by the adults' response: i.e., the adult is the interpretant (in Peirce's sense) of the infant's behaviour. [39]

We are dealing with a proto-semiosis or primary semiosis here, because there is still no semiotic substitution as in higher forms of semiosis. It is a semiosis without any sign, as the author explains:

> Paralleling developmental psychologists' notion of primary intersubjectivity is what one might call primary semiosis: i.e., the sum total of these phenomena. Primary semiosis covers all cases where meaning is co-constructed by actors in praesentia, where one does not yet have clear evidence of semiotic substitution (something standing for something else) as is the case in fully developed semiosis. (ibid.)

These assessments of the symbolic function of reinforced learning lead us to consider the sensory cortices' cells as an encoding of environmental stimuli deformed by the organism that is the source of pregnancies. Cells should therefore be considered memories of pregnant forms. Their function is semiotic in the moment when they are interfaces mediating between the salient forms of the world and the pregnancies of the organism. Considering that with 'einforced learning' we do not only mean Pavlovian conditioning but a learning "mediated" an interpretant, the neural morphologies that are obtained are already the result of a primary semiosis. Obviously, this is only a first level of stratification, which articulates forms of signification that will become progressively more complex, and which allows for the deployment of the interpretation planes.

If we look now at this proto-semiotic process from the neural point of view, meaning at the level of (part of) the problematic conditions of the production of phenomena, we see that this process implies the engagement of embodied plasticity that reshapes the neural assemblage by deforming symmetric connectivity and changing the geometric phase space of the assemblage. In this way, group symmetries that were at the centre of neurogeometrical theories are weakened for low level cortex and completely lost for higher levels of the brain.

But embodied plasticity also unveils a more radical heterogeneity of the brain structures related to affective, sexual and emotional process. This heterogeneity is linked to synapses that drive a plethora of neurotransmitters, neuromodulators and neurochemicals that are involved in all pregnant process. The relation between neurochemical circuits and affective cognition has been deeply studied by the spinozist neuroscientist Antonio Damasio and we refer the interested reader to the celebrated "Descartes' Error: Emotion, Reason, and the Human Brain" [11]. For our purposes, it is sufficient to outline that these synapses engender their own specific dynamics, thus inducing a variety of differential operators that change from point to point.

The virtual here is different not only from the fixed virtual of mathematical physics but also from the conception of the virtual that allows for the geometrical change of phase space. The virtual in this case is heterogeneous in both geometry and dynamics, allowing a true heterogenesis. From a dynamical point of view, this opens the differential towards sense-making, where the differential becomes plastic, embodied, situated as well as historically and socially determined. This virtuality embodies the conditions of deployment of a perception process that is already semiotic, since it implies the co-constitution of forms of the body and forms of the world as in the Merleau-Pontian tradition, in stark contrast to information processing in which there is only the constitution of saliences empty of meaning. While sense-making can rightfully be called a heterogenesis, information processing remains thoroughly homogenetic.

We will come back later to the topic of the genesis of primary semiosis and more structured semiolinguistic systems by investigating the emergence of the fluxes of expression and content as the actualization of heterogeneous assemblages.

5.5.2 Extended Cognition

Contemporary tendencies in the cognitive sciences consider thought and cognition not as localized in the brain but as extended in some way. For example, cognition can be conceived of as extended in the body-brain system (embodied cognition), as in the theory of enactive perception of Alva Noë (A. Noë, 2004). More generally, however, extended cognition should be taken to mean distributed between humans and non-humans, where thought emerges as a mediated process that makes the individual into one of the involved agents, rather than its organizing principle. These operate under the name of "embedded cognition", "extended mind", or "distributed cognition". Under these assumptions, cognition cannot be localized in the individual, but is constitutively distributed in a multiplicity of instances where individuals and their mental activity are only one of the constitutive dimensions.

> Memory does not reside in the folds of individual brains; rather, memory is the enfoldings of space-time-matter written into the universe, or better, the enfolded articulations of the universe in its mattering. Memory is not a record of a fixed past that can ever be fully or simply erased, written over, or recovered (that is, taken away or taken back into one's possession, as if it were a thing that can be owned). And remembering is not a replay of a string of moments, but an enlivening and reconfiguring of past and future that is larger than any individual. ([2]: ix)

For example, Hutchins [21] supports the idea that cognition has to be studied "in the wild", rather then confined in laboratories. In this condition, at least three interesting kinds of distributed cognitive process become apparent: cognitive processes may be distributed across the members of a social group, cognitive processes may be distributed in the sense that the operation of the cognitive system involves coordination between internal and external (material or environmental) structure, and processes may be distributed through time in such a way that the products of earlier events can transform the nature of later events.

These extended dimensions that are added to embodied cognition require intensive heterogeneity, which in turn implies a variety of virtual elements including, at the neural level, mirror neurons, which are at the centre of various theories. For example, Simulation Theory [18] is not only a neurocognitive theory but also a theory of social cognition, that is, a theory of how it is we understand others' actions, basic intentions, emotions, and sensations. In this setting, the claim is that "the fundamental mechanism that allows us a direct experiential grasp of the mind of others is not conceptual reasoning but direct simulation of the observed events through the mirror mechanism." [18]. But neuroscientific data concerning mirror neurons seem to be completely compatible with a number of other cognitive and social theories as well, such as the theory of narrative practices [22], the theory of interaction [17], the theory of enactive perception (Noë 2004). Particularly Claudio Paolucci in his book "Cognitive Semiotics" [28] propose a radical enactivist account of social cognition based on the heterogenous ontogenetic process that leads to mind reading through joint attention, semiotic competence, deception skills, pretend play, and language acquisition.

Anyway, more or less explicitly, all theories supporting distributed processes involve heterogeneous intensive assemblages. Eventually, the assemblages can be distributed among members of a social group implementing intensive constellations that are social, taking into account intersubjectivity, and also cultural, implying artefacts, texts and different semiotic systems. In 4E cognition, the virtual becomes then embodied, embedded, enacted, extended, and heterogenesis becomes the process of composition and actualization of the constellations of intensive elements involved in the process.[2]

But we must insist here that *heterogenous assemblage* does not mean *network*. That all becoming is nothing more than a communication game between agents who would be nodes of a network seems to us at least implausible or worse, uninteresting. Already interpreting the Deleuzian assemblage as a network (and how many readers have fallen into this trap) does not do justice to the indeterminacy and inaccuracy of material concatenations. Certainly the idea to allocating agency in heterogeneous associations of human and non-human is very strong and by no means trivial [27]. But it cannot be achieved with actants that communicate with each other through connective codes. If we look for symmetric anthropology between nature and culture we cannot avoid considering the individuation of agents, beginning with the continuous magma of forces and intensions that combine human and non-human, organic and inorganic elements. If there is an agency it can only be heterogenetic, where the concept of actant or agent is modified in the sense of a "magmatic", "uncertain", and "undetermined" agency that is always individuated by differential processes with their charge of uncertainty, non-commutativity and sensitivity to material conditions.

5.6 Imagination and Insight

Plasticity is a modality of the composition of operators which we could define as "passive" since the intensive elements and their combinations are induced by stimuli, affects and the social context. In Spinozian terms, we could say that plasticity has to do with a certain type of passive affects: passions. Let's see now instead how the virtual can be actively composed through a creative act, which for Spinoza corresponds to the affect referred to the action.

Imagination and insight occur when a problem is solved and can take the form of the comprehension of a joke or metaphor, the recognition of an ambiguous percept,

[2] About the relation between embodiment and artefacts see also "Corps Mutants" of Tiziana Villani [38].

or the realization that a project is feasible. They correspond to the "aha moment" in which illumination takes place.

What does it mean to have an idea?

> What happens when you say: "Hey, I have an idea?" Because, on the one hand, everyone knows that having an idea is a rare event, it is a kind of celebration, not very common. And then, on the other hand, having an idea is not something general. No one has an idea in general. ([12]:312).

Insight is often defined as a sudden change in the formation of a concept or other type of knowledge representation, often leading to the solution of a problem. These changes are thought to have certain attributes. For example, insights are frequently accompanied by a burst of emotion, including a highly positive surprise at either the content or manner of the realization. In contrast, analytic solutions are not typically accompanied by an emotional response except perhaps for a sense of satisfaction resulting from completing the task.

Another feature is that insights often break an impasse or mental block produced because a solver initially fixated on an incorrect solution strategy or strong but ultimately unhelpful associations of a problem. The breaking of an impasse is accompanied by the reinterpretation or restructuring of a problem to reveal a new, often simple, solution or solution strategy.

Specifically, we define insight as any sudden comprehension, realization, or problem solution that involves a reorganization of the elements of a mental representation of a stimulus, situation, or event to yield a nonobvious or nondominant interpretation. Insights are not confined to any particular domain of understanding, but we do not include all sudden realizations within this definition.

At the neurocognitive level, what we are witnessing in the case of insight is the activation of new configurations of neural groups and the integration of the signal on the concatenation of the circuits. So, there is a kind of common synchronization or vibration that is created, as we have seen for the constitution of percept. This would be a common individuation mechanism. Looking at the oscillatory activity of the brain, at the moment when people solve problems by insight, relative to solving identical problems by analytic processing, EEG shows a burst of high-frequency (gamma-band) EEG activity over the right temporal lobe, and fMRI shows a corresponding change in blood flow in the medial aspect of the right anterior superior temporal gyrus [26].

Facing the solution of a problem, a preliminary solution strategy could consist in considering the ideas that had been entertained to this point. It can happen that different aspects of the problem have been developed independently. From the morphodynamical perspective, this means that different assemblages have been already composed. The breaking of an impasse could thus be represented by a recomposition of these assemblages in such a way that new conjunctions are put in place, which leads in turn to a new metric in the assemblages which now will be completely reorganized in a new geometry. The actualized flux in this enlarged manifold can give rise to completely new solutions. Therefore, it is a question of modifying the assemblages in search of the configurations that are more powerful, that is, that are able to produce new phenomenal realities by resonance. We could perhaps refer to the sequence of eigenvalues of the resonance configuration. The more the first eigenvalues are large and the others are small, the more its eigenvector comprehends the whole configuration of the assemblage without remains, that is, it manages to integrate all the problematic elements.

Two concurrent processes intervene in the logic of imagination: the composition of differentials on the virtual plane and the condensation of singularities in its actualization. It is the sublime occasion of Kairos "which makes the solution explode like something abrupt, brutal and revolutionary." [13]. Hence, the Deleuzian idea that the creative act would take place as a composition of adjoined differential fields can be found in brain dynamics when different cortical assemblies find a way to join with one another. If we were in the presence of a network, we could refer to this as an act of 'connection'; as we've already considered, however, the wet brain is a differential neuromagma that goes far beyond the connective scheme of the network and contains all the uncertainties and indeterminacies of material concatenations. The conjunction of Riemannian and sub-Riemannian manifolds thus opens onto new spaces of possibilities within which heterogeneous dynamics bring out unprecedented configurations.

The differential neuromagma is the virtual of processes that are always a hybrid of the cognitive, perceptual, and affective in such a way that it is difficult if not impossible to separate the various components. This definition of the virtual is compatible with many theories of knowledge, but as a result of its intrinsic nature it tends to maintain a hybrid character, giving rise to actualized phenomena that are blocks of functions-percepts-affects-concepts on the line of a continuum of expressive modalities. The cognitive event, the sudden illumination that we could define as insight, can hardly be confined to a particular substance of expression. Rather, it is always a mixture of art, science, philosophy, and all cognitive modalities that cannot be reduced to specific disciplines.

Notice that the subject of the creative act is not the individual since in 4E cognition the assemblage is distributed among different agents, meaning that all the cognitive, affective and social dimensions are present. The creative act is then the endless process of collective enunciation.

Of course the collective practice of composition is not limited to the action of adding components to the assemblage. If the assemblage itself becomes the condition of conventional conformity, losing its imaginative character, the creative act will consist in the subtraction or substitution of planes with respect to existing configurations. For example, F. Guattari considers capitalism as a homogenetic process in that it produces an infinity of consumer products but it express only one pregnance or value: profit. The action of subtraction, hacking, and cutting elements of the process becomes an imaginative act capable of multiplying the occasions of the emergence of new forms of value. Or, following Franco Berardi [6], social morphogenesis consists in the disentanglement of social subjectivity from the automaton in all its forms. In this context, disentanglement means detachment from cognitive automatisms on one side ("We have to disentangle the autonomous life of words. Poets can do that. This is their job. They have to understand it. Their job is not for a small minority of men of letters. No. It's a job that they have to do in the streets, among the children, the congregants of Greta Thunberg." [6]) as well as from automatisms of financial capitalism in the other side ("Is there still a way to disentangle ourselves from the global financial order that shapes our politics as well as our imagination?" [7].) Far from a merely logocentric battle, all languages and practices are involved in the pursuit of disentanglement from conventional use and automatic deployments. It is in this possibility to compose by means of action, subtraction and hacking that the Deleuzian "I would prefer not to" becomes an imaginative act, subtracting itself both from the positivistic innovationism of the capitalist age and the dialectic machine of mutual oppositions.[3]

5.7 Metamorphosis

Any creative act implies a *metamorphosis* which latter term must be understood in a very specific sense: it involves not only the becoming of forms but also the becoming of the virtual that generates said forms. It will be worth pausing to consider how the concept of metamorphosis has been developed in different contexts—namely, in the context of dynamic structuralism, on the one hand, and in that of post-structural dynamics, on the other. Structural metamorphosis deals with a transformation of forms in a pre-given space of possibilities, whereas heterogenetic metamorphosis engages in addition a change of the phase space. This difference can be clarified by way of a comparative analysis of Goethe's and Ovid's metamorphoses, undertaken from a dynamic point of view.

Goethe deals with the metamorphosis of plants and with the becoming butterfly of the caterpillar. The shape of the butterfly is already inscribed in the space of the caterpillar's possibilities, in what we call phase space. In other words, becoming a butterfly is a controlled trajectory within the space of the caterpillar's possibilities. That is, it is a possible phase transformation of the dynamics which is already inscribed in the set of genetic/environmental possibilities. As we have seen, René Thom's works on this structural becoming are among the most important. In accordance with such a perspective on metamorphosis, it can be affirmed that a caterpillar will never become a wolf or a puddle of water, because it is not in its space of possibilities.

This is not true for poor Daphne in Ovid's Metamorphoses, who finds herself transformed into a laurel plant by the divine power of rivers. Becoming laurel can only occur through the assemblage of external dynamic elements that are not already contained within the space of possibility as it is initially given. The intervention of an exteriority, an unpredictable element external to the system, broadens the space of the possible. In this case, the exteriority in question is even embodied by a divine

[3] On the concept of automatism and disentanglement see also [5, 29].

intervention. In general, it will be considered an exteriority insofar as it is unknown to the system under consideration.

There is, therefore, in addition to a becoming of forms, a becoming of spaces of possibilities, which is developed on the basis of a twofold temporality. The time of individuation, or the axis of Khronos, and the time of the intensive mutation, that of Aion. It is a double becoming and double temporality, which we find inscribed in the verses of Ovid:

> Vix prece finita, torpor gravis occupat artus: mollia cinguntur tenui praecordia libro, in frondem crines, in ramos bracchia crescunt, pes modo tam velox pigris radicibus haeret, ora cacumen habet: remanet nitor unus in illa.

which can literally be translated as:

> She had just finished praying, that a heavy numbness invades her body: her delicate chest is wrapped in a thin bark, leaves grow on her hair, branches on her arms, her feet just before so fast become fixed in inert roots: she remains only beauty.

With the expression "on the hair leaves grow, on the arms branches" Ovid tells us that the individuation of the laurel, which is expressed through the process of plant growth (the arborescence process) is taking over the individuation of the nymph Daphne and that the intensive plane of becoming of Daphne is being transformed into the differential plane of the shrub. It is not a question here of pure transformations of form but of the passage from one process of individuation to another, that is, from one virtual plane to another. Rather than a morphing or transformation from inert form to inert form, it is a passage from becoming to becoming, from the becoming one form to the becoming of another, with the transformation taking place on the virtual level and the mutation being intensive, operational, differential. In short, what is transformed are the very laws of becoming, rather than forms directly.

Each of these two processes of individuation—that of the nymph Daphne and that of the laurel—continues to bring with it a sequence of differentiation, originally described textually by Ovid but taken up even more explicitly in the numerous iconographic representations of Apollo and Daphne. The nymph/laurel mutation appears so natural because the two individuations show the same sequence of differentiation: just as the arms of the nymph bifurcate from the chest, so in the laurel the branches bifurcate from the trunk; just as the leaves bifurcate from the branches, so the hands bifurcate from the arms; and the face bifurcates from the bust like the top of the tree from the trunk. It's as if the two morphogenetic processes arose directly from Turing's prepatterning [37]. What we find here are the four terms of the analogy, which latter is itself made possible through the double becoming of forms and of possibility spaces.

This embodied analogy is made possible by the two axes of the actualization of forms and the transformation of possibility spaces, regardless of the structures that are preserved. Metamorphosis therefore arises as an enigma to be solved, which is always reinvented and renewed.

In this sense, the metamorphoses of the ancient Latins or Greeks are heterogenesis or "becoming other" often in a literal sense as "Heteroiumena", which is the title that Nicandro gives to his book of transformations. This same expression "becoming other" will then be taken up again by Deleuze to mean the multiplicity of becoming of the subjectivation process and its continuum mutation in a becoming-woman, becoming-child, becoming-animal, vegetable or mineral, becoming-molecular of each species, becoming-particles.

References

1. Abbasi-Sureshjani, S., Favali, M., Citti, G., Sarti, A., ter Haar Romeny, B.M.: Curvature integration in a 5D kernel for extracting vessel connections in retinal images. IEEE Trans. Image Process. **27**(2), 606–621 (2018)
2. Barad, K.: Meeting the Universe Halfway. Press, Duke U (2007)
3. Barbieri, D., Citti, G., Cocci, G., Sarti, A.: A cortical-inspired geometry for contour perception and motion integration. Journal of Mathematical Imaging and Vision **49**(3), 511–529 (2013)
4. Baspinar, E., Sarti, A., & Citti, G.: A sub-Riemannian model of the visual cortex with frequency and phase. J. Math. Neurosci. **10**(11) (2020)
5. Berardi, F., Sarti, A.: RUN, forma, vita, ricombinazione, Mimesis (2008)
6. Berardi, F. Interview Le Monde Diplomatique, https://mondediplo.com/outsidein/an-interview-with-franco-bifo-berardi (2019)
7. Berardi, F.: And: Phenomenology of the end. MIT Press, London (2015)
8. Citti, G., Sarti, A.: A cortical based model of perceptual completion in the roto-translation space. Journal of Mathematical Imaging and Vision **24**(3), 307–326 (2006)
9. Citti, G., Sarti, A.: Neuromathematics of vision. Springer, Berlin (2014)
10. Citti, G., Sarti, A.: A Gauge Field model of modal completion. Journal of Mathematical Imaging and Vision **52**(2), 267–284 (2015)
11. Damasio, A. Descartes' Error: Emotion, Reason, and the Human Brain, Putnam Publishing (1994)
12. Deleuze, G. 2006 (1975-1995). Two Regimes of Madness, Texts and Interviews 1975-1995, Edited by David Lapoujade, MIT Press
13. Deleuze, G.: 1968. Columbia University Press, Difference and repetition, New York (1994)
14. Eco, U. 1979 (1975), A theory of semiotics. Vol. 217. Indiana University Press
15. Franceschiello, B., Mashtakov, A., Citti, G., Sarti, A.: Modelling of the Poggendorff illusion via sub-Riemannian geodesics in the roto-translation group. In: New trends in image analysis and processing—ICIAP 2017, pp. 37–47. Lecture Notes in Computer Science, 10590, Springer, Cham (2017)
16. Franceschiello, B., Sarti, A., Citti, G.: A neuromathematical model for geometrical optical illusions. J. Math. Imaging Vis **60**(1), 94–108 (2018)
17. Gallagher, S., Hutto, D.: Understanding others through primary interaction and narrativepractice. In: Zlatev, J., Racine, T., Sinha, C., Itkonen, E. (eds.) The shared mind: Perspectiveson Intersubjectivity, pp. 17–38. John Benjamins, Amsterdam (2008)
18. Gallese, V., Keysers, C., Rizzolatti, G.: A unifying view of the basis of social cognition. Trends in Cognitive Sciences. **8**(9), 396–403 (2004)
19. A.-J. Greimas, 1984. Semiotique figurative et sémiotique plastique, Acted Semiotiques, VI,60
20. Groupe μ, 1992. F. Edeline, J-M Klinkenberg, P. Minguet, Traité du signe visuel: pour une rethorique de l'image, Ed. Groupe μ, Seuil, Paris
21. Hutchins, E.: Cognition in the Wild. MIT Press, Cambridge, Massachusetts (1995)
22. Hutto, D.: Narrative practice and understanding reasons, In R. Menary (Ed.), Radical Enactivism. Focus on the philosophy of Daniel D. Hutto: 231-247. Amsterdam/Philadelphia:John Benjamins (2006)

23. Kanizsa, G.: Grammatica del vedere. Saggi su percezione e Gestalt, Il Mulino, Bologna (1997)
24. Klüver, H. (1926). Mescal visions and eidetic vision. American J. Psychol. **37**, 502–515
25. Koenderink, J., van Doorn, A.J.: Representation of local geometry in the visual field. Biol. Cybern. **55**, 367–375 (1987)
26. Kounios, J., Beeman, M.: The Cognitive Neuroscience of Insight. Annu Rev Psychol **65**, 71–93 (2014)
27. Latour, B.: 1991. Harvard University Press, We have never been modern (1993)
28. Paolucci, C.: A Radical Enactivist Approach to Social Cognition, in Antonino Pennisi and Alessandra Falzone (eds.), The Extended Theory of Cognitive Creativity: Interdisciplinary Approaches to Performativity. Springer Verlag: 59–74 (2020)
29. Pelgreffi, I.: Filosofia dell' automatismo. Verso un'etica della corporeità, Orthotes, Napoli-Salerno (2018)
30. Perona, P., Freeman, W. T.: A factorization approach to grouping, In H. Burkardt and B.Neumann, editors, Proc ECCV: 655–670 (1998)
31. Rizzolatti, G., Craighero, L.: The mirror-neuron system. Annu. Rev. Neurosci. **27**, 169–192 (2004)
32. Sarti, A., Citti, G., Petitot, J.: Functional geometry of the horizontal connectivity in the primary visual cortex. J. Physiol Paris 103(1–2), 37–45 (2009)
33. Sarti, A., Barbieri, D.: Neuromorphology of meaning. In: Compagno, Dario (ed.) Quantitative Semiotic Analysis, pp. 55–74. Springer, Berlin (2017)
34. Sarti, A., Piotrowski, D.: Individuation and semiogenesis: an interplay between geometric harmonics and structural morphodynamics. In: Sarti, A., Montanari, F., Galofaro, F. (eds.) Morphogenesis and Individuation, pp. 49–73. Springer, Berlin (2015)
35. Sarti, A., Citti, G., Petitot, J.: The symplectic structure of the primary visual cortex. Biol. Cybern. **98**, 33–48 (2008)
36. Thom, R.: Morfologia del semiotico, (Paolo Fabbri Editor), Meltemi Editore, Roma, 2006 (1981–1990)
37. Turing, A.M.: The chemical basis of morphogenesis. Philos. Trans. R. Soc. London. Ser. B, Biol. Sci. **237**(641), 37–72 (1952)
38. Villani, T.: Corps Mutants, Eterotopia France-rhizome (2019)
39. Violi, P.: How our bodies become us: embodiment, semiosis and intersubjectivity. J. Cogn. Semiot. IV(1), 57–75 (2009)
40. Warburg, A.: A lecture on serpent ritual. J. Warbg. Inst. **2**(4), 277–292 (1939), https://www.jstor.org/stable/pdf/750040.pdf
41. Weinberger, N.M.: New perspectives on the auditory cortex: learning and memory. In: Celesia, G.G., Hickok, G. (Eds.) Hanbook of Clinical Neurology, vol. 129, pp. 117–147 (2015)

Chapter 6
Expression and Semiogenesis

In the previous chapters, the question of heterogenesis, i.e. the question of an advent of actual forms on the basis of merely virtual resources, was taken up in two manners: on the one hand, we considered it in its foundations and with respect to its general principles, and, on the other hand, it was examined in closer detail according to certain more specific angles. In particular, one of the advances we've been able to make is to demonstrate that at the level of processes and phenomena of perception, the heterogenetic approach was able to bring to light the principles and the rationality of the salience-pregnance articulation. In this chapter, we propose to extend this exploration of the salience-pregnance articulation by placing it in a semiolinguistic light. Indeed, salience and pregnance constitute the facets of the originary expressive fact: as built according to the existential scenario of a co-constitution of a body and a world, which thus interpenetrate each other. Considered from a semiolinguistic point of view, the expressive phenomenon similarly proceeds from an interpenetration of the substances of expression and content. By way of semiogenesis, then, this expressive phenomenon is "polarised" and promoted to the rank of sign within a correlatively instituted semiolinguistic system. In order to account for this, we propose in these pages two further steps: first, to unveil in terms of heterogenesis the principle of cosubstantiality at the foundation of the semiolinguistic fact; and second, to go beyond the regime in which the expressing and the expressed, enveloped in one another, are taken to form a closed and undivided unity, in order to institute the order of the sign in which the signifier and the signified, although interdependent, hold the power to transgress the laws on which they are based for the purpose of renewal or adjustment.

© Springer Nature Switzerland AG 2022

A. Sarti et al., *Differential Heterogenesis*, Lecture Notes in Morphogenesis,
https://doi.org/10.1007/978-3-030-97797-9_6

6.1 Problematic Landscape

At the turn of the twentieth century, semiolinguistic science recognized the fragility of its views, even their extravagance, and as it became aware of the great difficulties it encountered in describing and explaining its object, semiolinguistics began, along various paths, to work on clarifying its principles, methods and objectives.

It was as much a question of delimiting the perimeter of its phenomena as it was of defining the forms and modalities of their determination.

In order to do this, it was necessary to specify what, in the tangle of empirical dimensions involved in any act of speech, properly constitutes a linguistic phenomenon; and correlatively, under which conceptual device and according to which procedure of analysis such phenomena and their observable functioning can be validly qualified. Such a procedure is at the risk of an overlap of the one and the other plane of determination, which would result in a descent into the epistemic circle (see below).

More generally, semiolinguistics, in order to go beyond the stage of free speech and risky speculation and to ensure that its views have objective value—in short, to escape insignificance and rise to the rank of authentic knowledge, semiolinguistics, thus, had to be constituted in relation to the principles which, in law and in practice, provide a general basis for the objectivity of discourse on the world, thus to be constituted in one form or another of empirical rationalism.

De facto, epistemological questioning occupies a central place in semiolinguistic reflection. There is no major undertaking in this field that does not devote a significant part of its work to explaining its foundations: its principles, methods, and criteria. Thus, to mention only three of the greatest: with Saussure the eminently epistemological question of the primacy of "points of view" on the object is posed from the outset; with Hjelmslev, it is the relational nature of the object of knowledge that is affirmed and assumed; and the Chomskian revolution will largely consist in endorsing the falsificationist epistemology, dominant in contemporary natural sciences.

In practice, the most thoughtful semiolinguistic endeavours will take up at their foundations epistemological categories and principles which go beyond them in that these categories and principles establish the objective value of empirical knowledge "in general" or the epistemic consistency of particular theoretical devices. In return, such principles and categories guarantee the credibility, until proven otherwise (falsifiability), of the theoretical dispositives that conform to them and of the determinations that these dispositives assign to the approached factualities.

But this virtuous conduct, which is undisputed to have contributed greatly to the recent advances in the sciences of signs and meaning, ultimately proved to be inappropriate in relation to its material. Simply because semiolinguistic phenomena are constituted in such a way that they fall outside the field of application of the principles and methods of knowledge in its "classical" epistemological format.

This is the case with the principle of refutation at the basis of Popperian epistemology, which it is accepted cannot be satisfied by semioinguistic theory. This inadequacy of the Popperian gnoseological device is due to the fact that semiolinguistic theories do not satisfy the architectural conditions of the refutable theories, i.e. to pair a "principal" theoretical component with an "auxiliary" component (phase space) which is worthy of an independent observation post. Now, as Milner explains, semiolinguistics is "scientia unica" ([25]: 131) in the sense that semiolinguistic phenomena do not come under any observation post detached from a semiolinguistic theory, i.e. they do not interest any science other than semiolinguistics. Semiolinguistics is therefore alone with its data: it produces and evaluates them, only to fall inevitably into the epistemic circle. In other words, since its theoretical concepts are at the beginning and at the end of its functioning (they produce the data they are intended to qualify), the judgements that semiolinguistic theory makes about its data will necessarily be found to be true. No refutation is possible.

The reader might object that this picture is painted too quickly: semiolinguistics would have at least two related observation posts—namely, phenomenology and neuroscience (cf. [31]). However, this picture is no less valid with regard to the logic of analysis and descriptive modalities actually implemented by a number of theoretical currents. Moreover, if phenomenology and neuroscience are proposed as observation posts for semiolinguistics, a closer look will reveal that they cannot fulfil this function (cf. [31]) for two reasons. On the one hand because the elaboration of empirical material for neurophysiological observation requires a qualification and a semiolinguistic organization of the data, which is therefore dependent on the theoretical choices submitted to the experimental test. And on the other hand because a phenomenological analysis of the sign highlights the mutual overlapping of the forms of its appearing and the forms of its conceptualization, in other words, the overlapping of the forms of its phenomenality and objectivity. In this case, the main espitemic separation of the orders of the sensible and the intelligible is outerpassed.

Another primordial notion of classical epistemology, which is incompatible with the constitutive regimes of semiolinguistic phenomena, is that of 'substance', on which it is appropriate to dwell for a while. This notion of "substance" can be taken in different senses, which are not mutually exclusive. First of all, from a Kantian perspective, there is substance as a category of understanding, thus endowed with an objective content and whose transcendental meaning is given by its schema (principle of construction of the concept in the forms of intuition), namely permanence in time. This understanding of substance thus contains the idea of a sustenance in itself, of a kind of eternal existence, and in this it is in line with the notion of substance as conceived by Descartes.

As far as the Cartesian perspective is concerned, it is, as we know, dualistic, in that it distinguishes between two substances: substance-thought and substance-matter, which are postulated in coexistence. In addition, relating the body to extension, the Cartesian perspective attributes to materiality the same *partes extra partes* constitution that it attributes to space (of which, in the Kantian perspective, the analogous form in this respect is that of juxtaposition). Body and extension (matter and space) are therefore considered to share a common structure: each is made up of distinct parts linked by a relationship of exteriority. In this respect, we know the radicality of the Cartesian thesis which identifies the body, extension and space, a thesis adapted by Kant, according to which real extension presupposes and is anchored in the extension of space: « The Cartesian and Kantian tradition [...] turns spatial determinations into the very essence of the object and it shows existence *partes extra partes* and the spatial distribution to be the only possible sense of existence in itself. » ([21]: 149). At the most general level, such an idea of substance thus entails the idea of its divisibility into a plurality of distinct parts that are mutually external to each other. However, if thought and matter are considered to be of distinct essences, it remains the case that conceived as substance they share the above-mentioned characteristics: « [...] although they are without common determination, [they] are nevertheless unified to a certain extent in virtue of the fact that both are conceived as substances » (A. de Waelhens, Introduction to ([23] : xix)).

To avoid any misunderstanding about the notion of substance as it appears in various philosophical perspectives, we outline that the notion of substance is here considered as a plurality of parts in relations of exteriority. Therefore it is not to be understood as a category of understanding (the homonymous category), i.e. a property of an object "in general", but as a primary modality of reason in its work of elaborating knowledge. In Kantian terms, substance here designates an idea and not a concept. Thus, although devoid of objective content, it nevertheless has a transcendental value in that it underpins the movement of reason in the increase and unification of knowledge. It is from precisely this angle that the notion of substance should be examined more closely.

First of all, it should be noted that the notion of substance in the epistemic sense we've begun to consider correspond to that of "matter" (of a phenomenon) as Kant presents it in his transcendental aesthetic, namely as "diversity of sensation" (or in kantian terms, a "manifold of sensation".[1] Substance, then, designates a homogenous and actual diversity—homogenous in that its units belong to a common measure, and, thus subject to relationships, can be linked under the unity of a concept (as a principle of synthesis), and actual, in that the substance relates the order of an effective existence. But this overlapping is not without inducing ambivalences that must be guarded against. To this end, it is necessary to clarify the concepts mobilized here—specifically, those of substance, matter and form.

Starting with the notion of form, we shall observe that, from an epistemic perspective, i.e. a perspective that is interested in knowledge from the point of view of its nature and production, and not from the point of view of its empirical, universal or particular, contents, the notion of substance has as its obligatory complement that of form, understood in the broad sense as a correlative of an act of knowledge. Let us consider this more closely.

As Hjelmslev insists, knowledge is not interested in a hypothetical "substance in itself" but in the relationships that occur within it: sound reasoning "[...] is opposed to any hypothesis that states or presupposes the existence of facts that logically precede the relationships that bring them together" ([13]: 32, our translation). In other words,

---

[1] « The effect of an object on the capacity for representation, insofar as we are affected by it, is sensation. [...] I call that in the [phenomenon] which corresponds to sensation its matter, but that which allows the manifold of [phenomenon] to be ordered in certain relations I call the form of [the phenomenon].» ([18]: 172-173).

science "[...] denies the scientific existence of an absolute substance, or of a reality that is independent of the relationships" (ibid., our translation). And since "[...] the only way to know (describe, understand) an object is to know its functions [here in the sense of relations]" ([14]: 76, our translation), all science "[...] has as its goal the knowledge not of individual objects but of functions" (ibid., our translation). In short, « the "objects of naive realism are [...] nothing but intersections of bundles of [relationships] » (Hjelmslev, 1969 (1966): 23), only the latter have a scientific reality, i.e. are accessible to knowledge.

Form, however, is not pure abstraction, as if residing in the sky of ideas, it also has a concrete dimension. Such an empirical realization of form is then called "substance". Substance can thus be understood as the tangible manifestation of form, the mode of the effective giveness of a form for an operation of knowledge which aims at recognizing it in a substance that delivers it.

Two consequences derive from this.

First, « [...] what from one point of view is "substance" is from another point of view "form" » (Hjelmslev, 1969 (1966): 23). In other words, form and substance are correlative terms, precisely in the sense that substance constitutes the as yet unanalysed data, either originally or as a complement to form, at a certain level of analysis or according to a certain prism of analysis. In fact, at a certain stage of its application, the procedure of analysis, which therefore consists in identifying a form distributed over different hierarchical levels, can consider its object either as analysed, therefore as form, or as still to be analysed, therefore as substance. Moreover, if the application of a prism of analysis which is correlative of a certain type of object (for example, physical, social, phenomenological...) leads to a form, this form always leaves a "residue" which can then be grasped from another angle of objectivity, and which therefore constitutes in this sense a substance. All this "[...] amounts to saying that in this general sense, 'form' and 'substance' are relative terms, not absolute terms" ([13]: 57, our translation).

Second, as a hypothetical receptacle of the form, the substance must also be understood as necessarily formless; as such, it bears, in Hjelmslevian terminology, the name "matter". In this light, it is a homogeneous diversity of actual atomic units, undifferentiated and mutually unbound.

To remove any ambiguity, let us recall the exact terms of this problematic of the relations between form, substance and matter.

According to the Hjelmslevian perspective, form and substance are defined as follows: form is « the constant in a manifestation » (Hjelmslev, 1969 (1966): 134), and substance is « the variable in a manifestation » (ibid.), where constant and variable designate, respectively, the necessary and contingent poles of a (unilateral) relationship of dependency, and where manifestation (D20) is defined as a « selection (co-presence relationship between constant and variable) between hierarchies and between derivatives of different hiérarchies » (ibid.). This essentially amounts to saying that substance is a relational structure whose existence is conditioned by a form, but without reciprocity—that is, the existence of form is not conditioned by substance. As for matter, it is defined as « [...] a class of variables which manifests at least two chains in at least two syntagmatics, and at least two paradigms in at least two paradigmatics » ([14]: 98, our translation). Let us neglect the second part (from "two chains...") of this definition, which is not essential for our purposes, and remember that in this definition matter designates a class of variables, and that matter, therefore, is to be understood as a collection of functional units. We can note, in addition, that these units all enter into the same relationship and, therefore, are mutually indistinguishable. The relation in question, furthermore, is a constellation—that is, a function involving two variables. The units making up matter are thus mutually unbound: they do not contract any kind of interdependence. Since knowledge is only interested in relations of dependence, matter is therefore inaccessible to knowledge: "for the aim of science is always to register cohesions, and if an object only presents the possibility of registering constellations [i.e. indépendancies] or absences of function, exact treatment is no longer possible" (Hjelmslev, 1969 (1966): 83).

To summarize: On the one hand, matter designates a homogeneous diversity, in that its units are identical and indistinguishable, thus sharing a common nature and being predisposed to forming a system; and yet it also designates an unorganized diversity, in that its units are mutually unbound. On the other hand, matter manifests a form, in the sense that matter is capable of instantiating it. Matter thus designates an actual diversity, that is an empirical amorphous set of units. Finally, the formed matter is called substance, which designates the unanalysed empirical phenomenon of which the procedure of recognition specific to each science will undertake to reveal the constitutive form.

Once this conceptual clarification has been accomplished, we will allow ourselves to name substance what Hjelmslev and Kant call matter. The reason is that matter is outside the field of knowledge: matter fundamentally expresses an epistemic hypothesis—an Idea in the Kantian sense—which is that of a substratum in itself unknowable (for the above-mentioned reasons) but presenting the characteristics of homogeneity and of actuality; homogeneity as the counterpart of its aptitude to receive a form and actuality in the sense that through matter the form acquires an empirical existence. Objectively speaking, only substance has a place in knowledge: matter is a background which is a necessity of a rational attitude and not of the forms of empirical knowledge, it therefore has a transcendental value but no objective content.

After this long digression aimed at clarifying the concepts of form, matter, and substance, we can resume the examination of the resistance that the semiolinguistic fact opposes to an analysis within the framework of a classical epistemology.

To the pitfall of self-consistency previously mentioned, we must now add that of ontological inadequacy. For—this is a fact to which we shall return in detail later—while substance is constituted on the basis of relations *partes extra partes*, semioinguistic facts are fundamentally elaborated according to relations of interiority, that is to say, in their generic scheme, in the mode of an "interpenetration" (or "reciprocal incorporation") of substances of content and expression, in other words, a "cosubstantiality" of the sensible and the intelligible. This situation, for all the reasons previously explained, is obviously inconceivable: inappropriate and inaccessible to thought, if not metaphorically. And even if, conceding the existence of separate spheres, we would like to conceive their junction and the formulas of their overlapping, we remain in the most perfect opacity. As Merleau-Ponty, for example, acknowledges when discussing the problematic interaction of the psychic and the physical in the phenomena of the phantom limb or anosognosia: Thus, « [...] we must attempt to understand how the psychical déterminants and the physiological conditions gear into each other [but] for the two series of conditions to be able to co-determine the phenomenon [...] they would require a single point of application or a common ground, and it is difficult to see what might serve as the common ground between "physiological facts" (which are in space) and "psychical facts" (which are nowhere) [...] » ([21]: 79). Such a "mixed" approach would bring together psychological and material causes in a way that is "fundamentally obscure": one cannot conceive of « [...] the incomprehensible encounter of two causalities » ([21]: 90).

The recognition of semiolinguistic phenomena, of which the "semiotic function" constitutes the fundamental feature, i.e. the essence character, requires us to take into consideration and conceptualise modes of relations of interiority, modes that are known to administer specific synthesis regimes (horizon synthesis vs. conceptual synthesis) and, at the same time, intermediary forms of presence and absence that make « [...] escape from the categories of the objective world where there is no middle ground between presence and absence » ([21]: 82), in other words, a world in which substances, in their empirical existence, take on their full and complete actuality. Such an undertaking is deemed impossible in view of the very conditions of all intelligibility, at least as elaborated within the framework of a "classical" épistémé.

To get out of this impasse, it is clear that it will be necessary to go beyond the classical conception of substance, as expressed through the previously exposed properties of homogeneity and actuality, and above all—and this is the crux of the matter—through its constitutive relations of exteriority (partes extra partes). It is therefore necessary to interrogate the notion of substance, and especially, to pose the question of its foundation—that is, what the classical conception of substance is based on. In other words, the necessary task is to question the substance (as a multiplicity) at its source in order to recognize its internal and formative principles. In order to overcome the obstructions inherent in the "classical" concept of substance, a theory of the genesis of substances must be developed which will make it possible to reveal new latitudes of functioning, and from which semiolinguistic phenomena could draw their intelligibility.

To this end, we shall proceed as follows: first of all, we shall return to the phenomenon of expressivity, in which the problematic "cosubstantiality" of the signifier and the signified manifests itself in all its fullness and obviousness. First, we will briefly recall the data, the reality and the difficulties of the problem, then we will set out the various ways in which semiolinguistics has developed in order to circumvent the problem of expressivity. In the following section, we will introduce the Merleau-Pontian solution, which theorises internal relations within the problematic framework of "solicitations", and which we will then see is affine to the heterogenetic perspective. Following this, to support the heterogenetic perspective, we will show that the problem of the first speech receives a solution within this framework. Finally, introducing a necessary semiogenetic process, we will see how morphodynamic structuralism, prolonging the heterogenetic dynamic, makes it possible to go beyond and theorise the « output » of expression, first by way of polarization, towards the sign, then by way of consumption and sedimentation, towards thought.

6.2 The Problem of Expression

6.2.1 The Expressive Phenomenon

The fact of expression consists in a sensible presentation of meaning, a tangible presence of significations. The fact of expression is striking due to its *paradoxical essence*. It is paradoxical because "[expression] announces a 'depth' which is *concealed* and which *reveals itself directly* within it" ([32]: 187, our emphasis). And it is paradoxical in essence because the contradictions which traverse it cannot be lifted without annihilating the object which actually proceeds from it. That is to say, the fact of expression, when it is a matter of considering it in its fully paradoxical essence, requires abandoning the distinctions belonging to classical epistemology and considering, with other forms of categoriality, new modes of the constitution of objects.

In fact, expression is an inconceivable mixture of sensibility and intelligibility, of intuition and understanding, of immediate and mediate knowledge, and, in fine, of presence and absence.

These difficulties concerning the expressive fact, and therefore the difficulties in clearly grasping and describing it univocally, can be avoided with the recourse to metaphorical language. For example, Taylor states that "Expression makes something *manifest* in embodying it" ([36]: 219, our emphasis). If the term "manifest", explained as being "directly available for all to see" (ibid.) so as to reinstate the full and immediate character of intuitive knowledge, belongs to the terminological and conceptual field of classical epistemology, that of "embodiment" is less firmly defined in conceptual terms. Nevertheless, this term still has the merit of expressing the indivisible unity of "expression and that which it expresses [...] and points towards the living, empathic presence within expression of that which is expressed." ([32]: 186-187).

But as pertains to approaching and to thinking about expressive facts, the difficulties are not only of a conceptual nature: They also concern the possibility of their observation. Indeed, the expressive fact, to put it as such, crumbles the moment when, by ceasing to be practiced, it acquires the status of an object, i.e. when it is "thematized". As Taylor observes: "An expression manifests something, but in an embodiment; and not any kind of manifesting-in-embodiment will do, but one that offers a physiognomic reading" ([35]: 78), that is, "Expression (...) involves what we might call direct manifestation, not leaning on an inference." (ibid.: 73). This is in contrast to readings that are "more analytical, geometric, or instrumental..." ([32]: 187). In fact, from the moment we steer away from the experienced singularity and presence of the expressive fact, from the moment we lose its immediate and always effective contact, be it by retaining what it expresses so as to inscribe it in thought or to submit it to reflection, or, conversely, by retaining the expression component in order to inscribe it within an act of interpretation, each time the expressive fact finds itself to be abolished. As observed by Merleau-Ponty (henceforth M.-P.) with regard to linguistic expression, the thoughts which accompany texts or discourses are not coextensive with them, but occur beyond the expressive fact, at the ulterior moment of a reflexive grasp or of a thematization, in which expression is then "fulfilled" ([22]: 40 & 59): "when a text is read in front of us [...] we do not have a thought on the margins of the text itself. The words occupy our entire mind [...] The end of the speech or of the text will be the lifting of a spell. It is then that thoughts about the speech or the text will be able to arise. Previously, the speech was improvised and the text was understood without a single thought; the sense was present everywhere, but nowhere was it posited for itself" (Merleau-Ponty, 2012 [21]: 185-186).

In short, the expressive fact does not suffer from the attention we place upon it, even less from our reflections concerning it. There is something of a "constitutive fragility" ([32]: 187) to expression. It exists only in the moment of its encounter and in its spontaneous exchanges, "But the moment people begin to reflect upon language instead of living it, they cannot see how language can have such power" ([22]: 8). Hence, the expressive fact, in that it assimilates its meaning with its manifestation, essentially signifies by weaving a world which is practiced and lived.

The fact of expression therefore pertains neither to a logic of communication, by virtue of which predefined contents are transmitted by means of a code, nor to a dialectic between interior and exterior, through which private internal states would be made public by its means. The fact of expression, in its irreducible essence, is simply that of the actual (sensible) presence of meaning. And it is this essential character which semiolinguistic analysis will attribute to sign phenomena—at least in what concerns the analyses conducted by Husserl, M.-P., and Saussure, which we will now revisit.

6.2.2 A Community of Views

Saussure, M.-P., and Husserl approach semiolinguistic facts using conceptual devices and angles of intelligibility which are greatly irreducible. Nevertheless, in the sort of liminal moment required for any theoretical investigation, when it is a matter of solely delimiting the empirical field of the discipline, they agree on the basics.

From the outset of their respective endeavours, and as if speaking through a single voice, Saussure and Husserl denounce the naïve conceptions according to which the sign would be an association between a tangible symbolic marking and a certain meaning. For both Saussure and Husserl, the sign does not rely on a distinction between sound and meaning. Such a dichotomy is fundamentally inappropriate for analysing the semiotic fact. In Writings, Saussure asserts that "it is wrong (and impracticable) to oppose form and meaning" ([4]: 17). This had already been anticipated in the Notes: "what is opposable to the physical sound, is [...] by no means the idea" (N9.2 in [12]:137), or again, he references the "obscurity and inanity of an opposition between the sign and the idea, between form and sense, or between the sign and meaning." (in [12]: 48). Husserl says as much: "It is usual to distinguish two things in regard to every expression: 1. The expression physically regarded (the sensible sign, the articulate sound-complex [...]); 2. A certain sequence of mental states [...] generally called the 'sense' or the 'meaning' of the expression [...]. But we shall see this notion to be mistaken" ([15]: 188). Likewise, for Husserl as well as for Saussure, it is necessary to distinguish the "true" sign, which has an indivisible nature, from the one resulting from a simple "assembly", that is, the "conventional" sign, as a correspondence between units of sound and a unit of meaning which are mutually foreign to each other with respect to their existence and to their principles of formation, and which therefore proceed from a logic of "naming-process" or of "communication" (ibid.: 189). Husserl (ibid.: 187) calls such signs "indicative" signs—these are the "commemorative" signs of the Stoics -, and he defines them as the articulation of two moments of consciousness: There is first a certain experience of consciousness, which is the perception of the symbolic marking, and, by virtue of its constituting function, the symbol reorients consciousness towards another content which is the thing, the idea, or the state of things to be communicated and of which the listener is to be informed. The "essence of indication" (ibid.: 184) thus resides in the fact that "certain objects or states of affairs of whose reality someone has actual knowledge indicate to him the reality of certain other objects or states of affairs, in the sense that his belief in the reality of the one is experienced [...] as motivating a belief or surmise in the reality of the other." (ibid.: 184). For Saussure, likewise, language is not organized in the manner of an index, that is, as a conventional reference of sound-units to meaning-units, each constituted within their own spheres: "The characteristic role of language with respect to thought is not to create a material phonic means for expressing ideas [i.e. Husserl's indicative sign]." ([3]: 182).

Opposing and contrasting with the conception of the sign as a simple assembly, Saussure and Husserl defend the principle of a sign of another nature, one which is unitary and integrated. It is then necessary to distinguish, following the respective terminologies employed by each, the "meaningful sign"[2] versus the indicative sign for Husserl, and the sound-idea grouping versus the signifier/signified unit for Saussure. And for both, it is a matter of acknowledging that which constitutes the essence of the "authentic" sign, that is, a sort of reciprocal incorporation of the sign's faces, which precludes soliciting the one without appealing to the other.

For Saussure, therefore, "the linguistic phenomenon always has two related sides, each deriving its values from the other" ([3]: 8); "one can neither divide sound from thought nor thought from sound; the division could be accomplished only abstractedly, and the result would be either pure psychology or pure phonology." (ibid.: 113). From the point of view of Husserl and in a similar manner, though already with an intentionalist inflection specific to his own system of questioning, the meaningful sign (which he also calls "expression" ([15]: 187) inscribes itself within a single moment of consciousness: The apprehension of unordered sensible data and

[2] Or "significant sign" (bedeutsam Zeichen) as opposed to the "indicative sign" (Anzeichen).

their elaboration into a sign phenomenon (the noetic moment) operates within a single intentional act, that is, the aim of an object of meaning. Therefore, meaningful signs signify in another respect than indicative signs do: Whereas the connection between the symbol and its meaning proceeds from an interpretative moment, which consists in redirecting the consciousness of the actual symbol towards the object of meaning, the connection of expression to meaning is intrinsic to it, this being its very principle of constitution: "the essence of an expression lies solely in its meaning" (ibid.: 199). In other words, whereas the symbol signifies in that it is "interpreted" (ibid.: 188), the expression signifies in the "strict sense" ([16]: 30) of the term: "The essential function of expression is to signify [...]; and this signifying function, inasmuch as it is essential, exists even when the expression indicates nothing" (ibid.).

In other words, the "true" signifier, which Husserl thus calls "expression", comprises in its phenomenal nature the orientation of consciousness towards a meaning, and it is this intentional directionality which shapes its appearance as a word-sign: "the 'meaning-intention' [...] characteristically marks off an expression from empty 'sound of words'" ([15]: 194). and it is therefore "In virtue of [intentional] acts [that] the expression is more than a merely sounded word" (ibid.:192). These views, in what they basically assert, are largely shared by M.-P., also according to whom the word has a meaning. This is clear in many passages in which he criticizes the intellectualist and empiricist approaches, each of which places meaning outside of the word, thus making the word into an empty shell: "these two theories [empiricist and intellectualist], however, concur in the claim that the word has no signification" ([21]: 182); "Thus, we move beyond intellectualism as much as empiricism through the simple observation that the word has a sense" (ibid.).

To say that "the word has a sense" is to say that meaning does not lie outside of the verbal fact, and, hence, that between the two there is no sequential link, be it of precedence (of the thought with respect to the word) or of inference (from the word to the thought). Thus, "For the speaker [...] speech does not translate a ready-made thought; rather, speech accomplishes thought. Even more so, it must be acknowledged that the person listening receives the thought from the speech itself" (ibid.: 183-184), and, moreover: "Speech is not the "sign" of thought, if by this we understand a phenomenon that announces another [...] in fact, [speech and thought] are enveloped in each other; sense is caught in speech, and speech is the external existence of sense" (ibid.: 187).

These considerations, which largely corroborate the positions of Husserl and of Saussure, receive an expressivist inflexion with M.-P. On the one hand, the word is approached as a sensible presence of meaning: "The word and speech [...] cease to be a manner of designating the object or the thought in order to become the presence of this thought in the sensible world, and not its clothing, but rather its emblem or its body" (ibid.), and, on the other hand, it is the impossibility of escaping the word, therefore of moving away from it, without abolishing it, which is thus emphasized: "If we push the research far enough, we find that language itself, in the end, says nothing other than itself, or that its sense is not separable from it" (ibid.: 194). So, as "the sense of a speech act can never in fact be delivered from its inherence in some speech" (ibid.: 196), it would thence be impossible to conceive it as such, in itself, without annihilating the fact of expression which it qualifies in part.

6.2.3 *Shared Difficulties*

We would indeed concede that there are a few difficulties in clearly conceiving the indivisible essence of the sign, difficulties which stem from the impossibility of jointly conceiving the unity of the sign and its dual nature as a signifier/signified composite. Indeed, if the unitary character of the sign is maintained, it will be identified with a fact of expression from which it will then inherit certain paradoxes, central among which is the impossibility of conceptualizing its two facets without annihilating it.

Conversely, if we approach the sign while considering its two-sidedness from the onset, we will then be confronted with their unthinkable reciprocal incorporation. Certainly, by putting the emphasis on the interpenetration of the moments of sound and of meaning, it had first been a question of highlighting some of the characteristics of the essence of the "true" sign in order to distinguish it from other sorts of semiotic factualities. But this manner of approaching and of conceiving the unity of the sign cannot be maintained. Indeed, one must concede that this participation of sound to meaning, and reciprocally, either pertains to ontological teratology (in sum, to an assumed "mystery") or is stricken with inconsistency.

Because, regarding this latter point, even while choosing to acknowledge a relation of reciprocal dependency between the signifier and the signified, the necessary character of such a dependency directs against the principle of an analysis along these terms and throws into question the relevance of the poles thus identified.

As recalled by Lo Piparo ([19]: 146 sq) in his comparative study of Stoic and Saussurean semiotics, "true signs" (designated as "meaningful" by Husserl and as "indicative" by the Stoics) are dual entities of which the second term "cannot be known in an autonomous manner" (ibid.), and they pertain to an "entirely relational ontology" in that the existence of the parts which constitute them is necessarily simultaneous. For Sextus Empiricus, "indicative signs fall under the typology of the "simultaneously relative [...], and here is the radical critique [he makes of them]: "The indicative sign does not exist" (Ibid.)—the reason being that, formulated in Saussurean terms, as a result of the fact that the signifier and the signified reciprocally condition one another, both in terms of their identities and their existence, this very manner of dividing the sign into two constitutive parts is in fact inconsequent and sterile. Particularly, when term A comprises term B in a constitutive manner, the inductive directionality (if A then B), at the foundation of Stoic semiotics, has no more reason to be.

Likewise for Hjelmslev: Concerning the relation of interdependence which, namely, reunites the planes of expression and of content, Jørgensen & Stjernfelt assert that it is "empty [in terms of heuristics] in the Hjelmslevian interpretation, precisely because it is relational in such a consequent manner: It welds the two terms together, to a point where they become inseparable [...] and if two terms always appear together, it becomes impossible to separate them at their own level" ([17]: 90).

The difficulties to which the theory of the sign is confronted are not only of a logical or of a conceptual nature: They can also be bolstered by empirical observations—although in such a case we have trouble distinguishing whether the difficulties registered are inherent to the observed fact or if they are the consequence of the system of observation by virtue of which the facts are approached and qualified. Thus, taking a more empirical stance, Tamba-Mecz identifies several facts which appear to contradict the principle of a cosubstantiality between signifier and signified: "Other experience data appear to shake these first convictions: [translation, paraphrase, synonymy] demonstrate the possibility of exchanging verbal signifieds considered to be equivalent, though they may be configured by dissimilar signifiers; or [homonymy and polysemy] [...] in short, being indissolubly linked by the formulation of meaning, verbal forms and meanings may nonetheless be 'unjoined' by means of analysis" ([34]: 37). Moreover, "the indefectible union of verbal signifiers and signifieds is contradicted by the exchanges between signifieds of all orders." (ibid.: 3).

Phenomenology is not outdone by this. The analysis of the sign proposed in Husserl's first Logical Investigation encounters the same difficulty as the one which was diagnosed by Sextus Empiricus when it was question of the "simultaneously relative": The signifier and the signified configure themselves and mutually presuppose one another to the point that the appearance of the signifier and signified fully overlap in a signitive phenomenality which is then logically indivisible.

Indeed, we have seen that, beginning with the first Logical Investigation, Husserl distinguishes two regimes of significance, respectively, at work in indicative and in meaningful signs. We have also seen that the orientation towards an object of content constitutes the essential character of the meaningful sign: Whereas the indicative sign administers a correspondence between two experiences of consciousness constituted outside of one another, the meaningful sign incorporates in its appearing the mode of a consciential aim of signification.

Now, this conception of the "meaningful" sign is unsatisfactory because if such was indeed the case, then the appearing of the sign would fully inscribe itself within the appearing of meaning, as an object that is a target of the meaning-intention. Indeed, if the phenomenal identity of the sign found itself to be integrally configured by the sole consciential aim towards an object of signification, then the sign would never present itself otherwise than as meaning or as an integral (indissociable) part of a meaning: It would, exclusion being made of any other (manifest) character, be constitutive of a "pure" presentation of meaning "in itself"-thus obliterating the concrete dimension of the signifier. We can illustrate this conjuncture by means of an analogy with spatial perception. As the immanent adumbrating (the set of discontinuities which a spatial body projects upon the retinal surface, discontinuities which are indeed perceived and present within consciousness) finds itself to be spatialized and presented (to consciousness) as the apparent contour of a three-dimensional object, so would the medium (graphical or vocal) of a signifier be semiotized into an intrinsic component of the intended object of meaning. The medium would then disappear from consciousness as regards its immanent sensible characters-as is the case with a perceptual representation where "an experienced complex of sensations gets informed by a certain act-character [...] the perceived object appears, while the sensational complex is as little perceived as is the act in which the perceived object is as such constituted" ([15]: 214). But it appears that it is precisely the contrary which is revealed by phenomenological analysis: The concrete characters of the signifier persist, albeit in an altered form, in the perception of the sign.

Since it is a matter of approaching the reality of the sign, its contents, and its internal forms, and since the difficulties that arise in this pursuit are at once so numerous and so considerable, recourse to metaphor is frequent. Thus, we will encounter mentions of the "cosubstantiality" of the faces of the sign, of their "fusion", of their "reciprocal assimilation", and of their mutual "incorporation." For example, Benveniste states that "there is such a close symbiosis between them [the concept and the sound image] that the concept [...] is like the soul of the sound image" ([1]: 45) or: "The signifier and the signified [...] together make up the ensemble as the embodier and the embodiment [...]. This cosubstantiality of the signifier and the signified [etc.]" (ibid.). Such metaphors, however, do not resolve the problem, which remains in full.

Saussure, on the other hand, albeit without specifically discussing the difficulties of the dual and indivisible sign, promptly abandons the conception of a "consubstantiality" of the two faces of the sign, recognizing its unintelligibility: If "neither are thoughts given material form nor are sounds transformed into mental entities" ([3]: 112), it is because the sign is nothing but a "side effect." The sign is the functional consequence of a superior systemic reason (language) which operates by correlating relations of reciprocal delimitation in the substances, respectively, of expression and of content, in order to dually institute units (in these substances): "The characteristic role of language with respect to thought is not to create a material phonic means for expressing ideas but to serve as a link between thought and sound, under conditions that of necessity bring about the reciprocal delimitations of units" (ibid.). And, correlatively: "Language works out its units while taking shape between two shapeless masses" (ibid.). We know that this level of elaboration of Saussurean thought (which Hjelmslev regarded with severity) is not without weaknesses, and that it is in the framework of a theory of value, developed in the Third Course, that he will find the key for discarding once and for all a "great illusion", inasmuch as "to consider a term as simply the union of a certain sound with a certain concept is grossly misleading" (ibid.: 113).

6.2.4 *The Semiotic Function*

In contemporary semiotic science, the essential character of the expressive fact, namely, the reciprocal incorporation of the sensible and of the intelligible, is approached and qualified in terms of semiotic function. Precisely, in Hjelmslev's theoretical system, the semiotic function is defined as a relation of interdependence between the planes of expression and content (« [the semiotic function] is in itself a solidarity. Expression and content are solidary—they necessarily presuppose each other » (L. Hjelmslev 1969 (1966): 48). But although it has the status of a fully fledged theoretical concept, the semiotic function remains a blind point in semiolinguistic knowledge.

Even if some of the operative forms of the semiotic function have been clearly recognized (for instance, the commutation operation or its role in infinite dynamical semiosis, see below), its complete and deep comprehension has not yet been achieved. In fact, when the semiotic function is explicitly taken into account in a theoretical framework, it is generally reduced to surface operating schemes, which, although they do proceed from it, relegate its essential part to a more or less theorised background. And finally, the semiotic function is excluded from the scope of an explicit semiolinguistic knowledge.[3]

To illustrate such difficulties, inherent to the very theorization of the semiotic function, and to propose a possible way to overcome them, we will quickly examine a series of theoretical apparatuses that are either historically important or currently prevalent.

6.2.5 Saussure

As we have seen (cf. above), if at the beginning of the Course Saussure suggests a kind of merging (blending) of the signifier and the signified, he will quickly abandon this conception in favour of a functional architecture in which the unity of the signifier and the signified is partially rebuilt.

Through a mathematical formulation of the topological and dynamic intuitions at the heart of Saussurian thought, we show (cf. infra) that between the signifier and the signified there is precisely an asymmetrical relationship: the existence of the signified presupposes the signifier in that the latter, in the functional position of a control parameter, determines the actualization of differential relationships in the substance of content, relationships that precisely institute the signified.

Finally, the undivided unity of the signifier and the signified will have been dissolved (albeit partially) in the functional system of language. The originary awareness of the expressive fact, which is a matter of perception, is then overcome in favour of a characterization, which pertains to knowledge, where the signifier and the signified

---

[3] Commonly, semiolinguistic theories only record and exploit the correlations between forms and meanings, thus without approaching their internal principles. The undivided unit of the sign is thus generally related to a coupling, either logical or dynamic, of a signifier and a signified that are constituted independently one from the other. Thus, for instance, in construction grammars the integrated unit of the word form and of its meaning results from an associating and storage process ("entrenchment") of a routine of co-actualization (of form and meaning) based on the reiteration of co-occurrences, and not from an « interpenetration » (of form and meaning). Word form and meaning are thus elaborated beforehand and independently one from the other.

are the functional components of a systemic totality devoted to the production of meaning. This means, in semiogenetic terms, that the Saussurian sign constitutes an "overcoming" of the expressive fact, an "overcoming" which, through the effect of a polarization of expression into a signifier and a signified, gives the speakers the capacity to modulate and adjust new meanings constantly. We will come back to this in detail in the pages devoted to the morphodynamics of the sign.

6.2.6 *Hjelmslev*

In Hjelmslev's glossematic theory, the process is quite similar but with the merit of clarity. In fact, if the undivided unity of expression and content is again placed in a theoretical background, the process is now explicitly theorized: in the glossematic apparatus, the set of relationships on which the semiolinguistic objectivity is built is located at a level of analysis that is hierarchically below the level of the connection between the planes of expression and content. In this way, the semiotic function is located outside the scope of linguistic knowledge and appears to be indeterminable.

More precisely, the relationship between the planes of expression and content is conceived in terms of "solidarity": "the semiotic function [...] which unites the plane of content with that of expression [...] is a relation « and ... and » [i.e. a syntagmatic link], since the two planes are coexistent and not alternative, [and] between the two planes there is interdependence, since they are complementary" ([13]: 159, our translation). But this "solidarity" is not a matter of form, because "the distinction between content and expression is the first crossroads [of the analysis], that of form and substance the second, and the distinction of form and substance is therefore subordinate to that between the planes" ([13]: 53, our translation). It follows that it is possible to speak of a form and a substance of expression or content, whereas "[...] it would be senseless, because it is inappropriate, to speak of a 'content of substance', a 'content of form', an 'expression of substance' or an 'expression of form'" ([13]: 53, our translation). Consequently, since the object of a knowledge is a form (cf. supra), the semiotic function escapes all knowledge.

Correlatively, the relations between expression and content units (for instance, between signifier and signified) are rebuilt on the ground of the commutation relationship, which is precisely defined as a « conjunction » (« both-and function », cf. (L. Hjelmslev, 1969 (1966): 36)) between « disjunctions » (« either-or function », cf. (L. Hjelmslev 1969 (1966): 36)) recorded in each plane. What we observe is that once again the primacy given to functional architecture is at the expense of the semiotic function, which is nonetheless recognized by Hjelmslev as the essential feature of all authentically semiolinguistic phenomena.

6.2.7 *Husserl*

The phenomenological perspective is not to be outdone: the analysis of the sign developed by Husserl has come up against serious obstacles, which it has not succeeded in overcoming. Since the problem of the undivided unity of the sign does not find an internal answer, it is ultimately by resorting to the superstructure of the attentional field of consciousness that the signifier and the signified recover a certain unity. Indeed, it should be recalled (cf. supra) that if, as the first Logical Investigation states, "the essence of an expression lies solely in its meaning" ([15]: 199), then phenomenological analysis refrains from recognizing, within the "sign phenomenon", the presence, however unmistakable, of a component that is simply sensitive and jointly given, albeit in a weakened mode, to the object of a signifying intention (the signified). To overcome this difficulty—which would be to preserve and assemble two intentional aims, one of perception and the other of signification, but at different levels of consciousness—Husserl will have recourse to the external structure of an attentional field (introduced in the fifth logical investigation and taken up again in Ideen and Lessons...), where the two aims are situated at distinct but interdependent levels. But in doing so, if the phenomenological description of signifiers and signifieds is partly acquired, it is the unity of the sign that is then lost, simply because the two aims are elaborated independently of each other, thus contravening the type of unity decreed by the semiotic function.

6.2.8 *Discussion*

These three approaches have been used jointly to recognize at their starting point the essential character of the semiotic function. However, in the developments offered by these three great theorists of the sign which aim at the determination of the signs in their functional composition and their connections to the other signs, the primary fact that semiotic function reports is in each case left out or exceeded. Everything proceeds as if, when it is matter of clarifying the functional regulations and the relational modalities that determine the semiotic phenomena in their empirical objectivity, the first condition of any semioticity, the essence of the semiotic fact, to resume Husserl, is placed in the background as an implicit foundation. It may well be considered unknowable, but it is always contiguous to the determinations that the theoretical devices in question deliver.

We then understand why many semiolinguistic approaches have been developed by treating the semiotic function only obliquely or indirectly. Such is particularly the case of the Peircean semiotic, which we will examine schematically below.

6.2.9 Peirce

As is well known, the core of the Peircean apparatus articulates three terms, one of which (the Object) is split, namely (i) the sign (or representamen), (ii) the interpretant (which is also a sign), and (iii) the Object, in which one will distinguish two aspects: (iii-a) one pertaining to the real world and called the dynamic (or dynamoïd) object, and the second (iii-b) pertaining to the semiotic system: the immediate object.[4]

We know that the sign, in its role of representamen, refers to the object it represents through the mediation of other signs, which then act as interpreters, and that it has the power to "trigger": "[the interpretant is] a sign which returns a representamen to its object" ([5]: 21-22, our translation), in this sense "[the interpretant] operates the mediation between the representamen (first) and the object (second)" ([8]: 40, our translation). The interpretant is thus the active principle of the semiosis in that it establishes the link between the representamen and what the representamen refers to. We know that this functional configuration opens an unlimited process of semiosis: the interpretant, as a sign, calls to others interpretants, and so on, endlessly.

We can now examine the relationship between signifier and signified within the framework of this theoretical device. In the Peircean apparatus, the role of signifier is clearly taken by the representamen, which is "the sign as it is presented and that the interpretant will [then] refer to the object it represents" ([5]: 23, our translation, we underline). Concerning the role of the signified, the case is more complex because the Peircean device is dynamic, and the content assigned to a sign is the asymptotic limit of an endless process of semiosis. Accordingly, depending on whether one is interested in a state of the signified corresponding to a given stage of the semiosis process or corresponding to the limit of the infinite semiosis, the role of the signified will be carried respectively by the interpretant or by the immediate object: "It seems natural to use the word meaning [signified] to denote the intended interpretant of a symbol" (C. S. Peirce, 1931-1935: §5.175) and elsewhere, "the complete Immediate Object is identified with the signified" [CP 2.293] » ([7]: 108, our translation). Whatever the option is, the one reporting the signified to the interpretant, the second to the dynamic object (i.e., the object that a series of interpretants gradually circumscribes), the semiotic principle is carried by the representamen insofar as this latter, in its quality as a sign, opens onto other signs, or, according to the canonical definition of the sign, "determines" the interpretants that contribute to configuring an immediate object. In this context, the semiotic function is moved onto a phenomenological plane: the representamen essentially implies an opening towards something other than itself, be it a sign or an immediate object, and this characteristic is expressed in the very moment of its giveness as a sign, because it configures its appearing. This is why we must recognize with Eco that the term *signified* seems « at once a semantic category and a category of the phenomenology of perception "([6]: 33): it is only

[4] « It is usual and proper to distinguish two Objects of a Sign, the Mediate without, and the Immediate within the Sign. [...] The Mediate Object is the Object outside of the Sign; I call it the Dynamoid Object. » (MS [R] L463, in [26]).

because I know that smoke means fire "[that I] am able to render the sensory datum meaningful, by seeing it as that smoke which can reveal fire" (ibid.).

We can thus draw the following provisional conclusion: in a manner similar to the theoretical frameworks previously discussed, the Peircean conception, which accounts for the (dynamical) elaboration of meaning and signs, sets the semiotic function on a phenomenological plane which escapes the functional determinations of the theoretical apparatus. However, it will be useful to further consider the Peircean device, particularly insofar as it was taken up by Eco and brought to bear upon an interpretation of the Hjelmslevian apparatus.

6.2.10 Eco/Hjelmslev

In the case of Eco, the question at hand is no longer primarily that of the sign, but rather that of the conceptual triad: form, substance, and matter (or purport). What is at stake here is the constitution of substances and, in the end, the possibility of understanding the internal principle of the semiotic function. Let us first recall the three notions in the Hjelmslevian apparatus (cf. supra).

The form is an ideal structure, specifically an abstract network of dependency relationships. When this form becomes incarnated and manifested, it is precisely denoted by the concept of substance. The third term, the purport, is related to the amorphous manifold that is modelled by the form when form is projected onto purport, producing substance.

In glossematic theory, the purport (or matter) is defined as an amorphous aggregate of independent and unitary atoms. In defining purport in such a way, Hjelmslev locates it at the boundaries of what can be known. Indeed, on the one hand, matter is located outside the field of knowledge, simply because knowledge concerns only "cohesive" relationships that do not belong to purport's units. On the other hand, the purport can be conceptualized since, because it is apt to receive forms, it must hold the qualities by which such an instantiation is possible. Thus, even if it is free of form, the purport is minimally formed (as a set of univocal and mutually untied atoms) to constitute the homogeneous soil for possible actualization of forms.

What Eco reconsiders is the idea that there would be two distinct purports, one of expression and the other of content, which would be understood respectively as the receptacles of the expression and content forms to produce the corresponding substances of expression and content. From Eco's point of view, what Hjelmslev calls purport corresponds to the Peircean dynamical object, simply because, like the dynamical object, the purport in the Hjelmslevian conception escapes all knowledge and constitutes a field to be "semiotized". Indeed, Peirce defines reality as "[...] the limit of what can be known, what would be known by an infinite semiotic practice" ([8]: 45, our translation) and considers the dynamical object as "[...] what the sign refers to in its existential singularity" ([5]: 66, our translation). The Hjelmslevian purport is similarly a manifold of singularities without any form or cohesion, and then it is located outside the field of any knowledge. Conceived in this way, the purports of expression and content cannot be distinguished one from another, since they are defined in the same way: they are amorphous aggregates, as untied punctual diversities, and do not hold any organizational characteristics that discriminate them. From this point on, we will gladly follow Eco's thesis, which "represents the continuum of the expression and the continuum of the contents as a same entity" ([7]: 80, our translation): "The matter, the continuum about which and through which signs speak, is always the same. It is the Dynamic Object that Peirce talked about [...]" ([6]: 44). As a consequence, too, "the continuum which one forms to express itself is the same one than that which one expresses" ([7]: 80, our translation).

At this point, we note that this theoretical adjustment, introducing the assimilation of the purport of expression and contents, is insufficient to enlighten the semiotic function, since this function is committed between the planes of expression and of contents through their articulations of substance/forms, and therefore does not imply in any way, other than in an atheoretical background (see above), the presence of the purport. Furthermore, this is clearly represented in Eco's diagrams: it is within the interior disc, subdivided into two half discs, one for expression, the other for content, that are constituted the units of form and substance, respectively, of expression and content, and that their connection (the semiotic function) is performed. The part between the exterior and interior circles, which thus represents the common purport, is not implied in the elaboration of the links between units of the expression and content planes, at least directly and formally. And it is on this latter point that the theoretical reconfiguration that Eco operates is essential.

6.2.11 Eco/Peirce

What is at stake now is the relationship between the dynamic object and the immediate object, i.e., the relationship of reality to what one expresses of it, or in other words, using Hjelmslevians terms, the relationship between the purport and the form. Concerning this point, on the side of glossematic, the question is clear: the purport constitutes a completely passive receptacle and is able to receive any semiotic formation. That is to say, the matter does not express by itself. The Peircean point of view is quite different.

As we have seen, "the immediate object is the mode of donation [i.e., the meaning as defined by Frege] of the dynamic object" ([7]: 108, our ranslation). But this mode of donation, which is a certain point of view of the object, is not, according to Peirce, arbitrary: it is not decided in the sign system but emanates from the dynamic object itself: "It is the dynamic object which determines the representamen to represent it through a certain point of view, the one of the immediate object" ([8]: 44, our translation). Then there is a first experience of the world (dynamic object) that originarily meets a universe of tensions, balances and constraints, waiting to be constituted as qualified phenomena, but that already orients "a certain point of view" of it. Thus, it can be said that "[...] it is under the pressure of the world (as a dynamic object) that the sign represents the world [...]" ([8]: 44, our translation) and that "[...] the immediate object reflects a meaning already implicit in the dynamic object" ([7]: 108, our translation). It will then be necessary to question the meaning and status of this implicit meaning, which is, in Merleau-Ponty's words, like a "preparation for the object".

This Peircean conception becomes even stronger when regarding the difficulties encountered by the converse positions developed in the Hjelmslevian apparatus (see further). From Peirce we will retain the conception of a first ensemble of solicitations, a first fabric of dubious impressions, an expectation of reactions and positioning, that gradually, in the way of individual experiences, take form, meaning and even a statute of object. Moreover, this option finds other supports when examining certain contemporary approaches to the semiotic function, notably the thesis of Fontanille.

6.2.12 *Fontanille*

From Fontanille's point of view,[5] and in agreement with recent developments in semiotics, semiosis is basically a matter of one's own body. In fact, after he has observed that "the body explicitly came back in semiotics", he continues that "the anchoring of semiosis (is) in sensible experience" (ibid., our translation). More precisely: "as soon as we wonder about the operation which joins together the two planes of a language, the body becomes essential [:] it [the body] has to be considered as the only instance common to the two faces [signifier/signified] or to the two planes [expression/content], and which can ground, guarantee and carry out their union in a meaning unit" ([11]: 13, our translation). The body is thus conceived as an "operator of semiosis" in different ways. First, the body takes part in the elaboration of sensible qualities of which it constitutes the praxical side: "each sensory apprehension is an apprehension of the motion, which accompanies, precedes or causes the motion, and which, consequently, is originarily a sensation of the flesh and of the body motion" ([10]: 9, our translation).

We see that at this first level of correlation between the sensory experience and body commitments, sensible qualities, are intrinsically meaningful, being praxical values. But it is only in subsequent operations (of conversion[6]) that significances attached to axiologic dimensions (note that axiology designates systems of values (ethical, logical, esthetical and more generally anthropological)) are processed and assigned. The body operates this time as an operator which, on each level on which

[5] Fontanille never misses an opportunity to underline the central role of the semiotic function: "let us take care of the fact that [analysis] respects the minimal constraint of a solidarity between expressions and contents" ([9]: 12, our translation).

[6] "[conversions] are operations which imply an epistemological subject equipped with a body, which perceives significant contents and which calculates and projects their values. For each change of level of pertinence, one can attribute the re-articulation of meanings to the activity of this sensitive and'embodied' operator: he perceives the meanings of a first level as tensions between categories, as graduated conflicts, and he draws from this perception new meanings, articulated as'positional values", on the next level of pertinence" ([11]: 14, our translation).

it operates, produces and projects new layers of meaning. What we observe on this second level and at the subsequent ones is that semiosis is conceived as a process that, by means of the power of the body and its affects, processes and "computes" new values of content to a plan of expression previously made up. In this perspective, there is no longer a semiotic function as we have introduced, but a process of semiotization through a reconfiguration and attribution of values of meaning. In Fontanille's view, we have to place ourselves at the stage in which bodily motions are accomplished correlatively with the installation of meaningful sensible qualities in order to see a genuine semiosis at work, which lacks at the levels of conversion (second and further levels).

We are not interested here in discussing the relevance of such an approach, since its interest and its efficiency have to be established elsewhere. We will just observe that the semiotic operation that is at its base and constitutes the first layer of expressivity remains obscure. Especially, we will observe that, set in that way, the problem of the semiotic function has been faced in the problematical framework of the "first" Merleau-Ponty problematical framework towards which we will quite naturally be redirected.

6.2.13 Conclusion

We hope that the overview we carried out of some of the major semiolinguistic theories, even if limited and schematic, can help identify some of the main obstructions to understanding the process of constitution of the semiotic function. More importantly, we hope it can suggest possible pathways towards the overcoming of these obstructions.

Accordingly, from the examination of glossematics, we will retain the deadlock induced by the reduction of purports to an aggregate of undifferentiated and homogeneous atomic units, in which, by way of a set of relationships, the purport would be informed in such a manner as to be productive of substances. From the analysis of the Peircean apparatus, we will retain the importance of conceiving an originary unspecified purport that exercises a "pressure" that impels the self-constitution of its meaning and its phenomenal forms. Finally, the approach of Fontanille, together with a general movement in semiolinguistic science, identifies the body as a "strange signifying machine" ([21]: 114) according to Merleau-Ponty's terms.

To deploy both this approach and the Peircean intuition of a dynamical object as experiential unspecified background "which questions the body", we will turn to Merleau-Ponty, who has produced some of the deepest reflections about meaningful morphologies and, more generally, about the semiotization of the world in relationship with a body. This will constitute our background to conceive the emergence of the semiotic function in terms of an heterogenesis, and thus, more generally, to overcome the paradoxes and aporias of « cosubstantiality ».

6.3 The Merleau-Pontian Solution: Toward Heterogenesis

6.3.1 The Problem of Solicitations

This section introduces the problematic of "solicitations." We shall show not only that this problematic sets up and explains the mode of "interior" relationships (as opposed to the "exterior" relationships described in section "Problematic landscape") at the basis of expressive facts, but also that it finds its accomplished formulation in a heterogenetic apparatus, within which, in particular, the relationships of "cosubstantiality" are elaborated and qualified.

In "The Structure of Behaviour" ([23]—henceforth SoB), a work preparing for Phenoménologie of Perception, Merleau-Ponty introduces a solution to the problem of expressivity, albeit in a lateral way. For Merleau-Ponty's aim in SoB is not so much to explain the expressive fact, in the sense of exposing its principles, laws and internal forms, but rather to investigate it in the double and crossed sense of establishing its obviousness and revealing its necessary genesis. In other words, the merleau-pontian approach in SoB establishes the expressive fact as much in its manifest characters as in its order of intelligibility, namely that of "internal" relations. To do this, SoB first examines the reign of life through the prism of experiments on physical and biochemical dimensions. It then appears that the explanatory modalities of the empirical sciences, i.e. essentially the reduction to more or less integrated and interacting causal chains, do not make it possible to account for observable behavioural regularities (or behavioural disintegrations), and that, by contrast, the behavioural patterns then highlighted find their rationality when they are considered from the angle of an elaboration of "internal" relations that integrate the organism into the world as it is perceived and integrate also the parts of the body into the living totality that they make up.

A few precautions should be taken at this stage : first of all, the resistance of phenomena of life, as well as those of expression and perception, to the explanatory categories of the empirical sciences is not a novelty—at least since Kant (third critique) who clearly distinguished the sphere of "material" phenomena whose objective value (the meaning of object) proceeds from the concepts of a transcendental understanding, and are therefore the object of determining judgements, versus these phenomena that carry a meaning in their own, i.e. whose meaning is immanent to them, and which, as such, lend themselves to reflective judgments. This boundary between the fields of determination and reflection—that is, between phenomena whose objectivity stems from a priori given categories, for example categories of cause or substance, versus those phenomena whose recognition is effected by means of interpretation—is well and truly established. The positioning of a class of phenomena on one side or another of this boundary is obviously a major epistemological issue, and any repositioning will also constitute a major scientific advance—for example the one accomplished by J. Petitot who, in the framework of a morphodynamic structuralism, was able to "[...] transform the 'supplement' *to* objectivity (which Kant was therefore appealing to reason) into a 'supplement' *of* objectivity, [thus] tilting an important part of what for Kant was an objet of reflective judgement to the side of determinant judgement" ([28]: 46, our translation).

Moreover, the theoretical systems and experimental data that Merleau-Ponty mobilizes in his demonstration are, to say the least, outdated (among others: reflex arc or conditioning theories). In the age of cognitive neurosciences, one may wonder about the interest of a demonstration which invalidates them and furthermore about the relevance of the conclusions which follow. The answer to these two objections is that Merleau-Ponty's objective goes far beyond the simple intention to refute: it is not so much a question of invalidating the theories under discussion as of grasping the facts that go beyond them—those facts which are revealed in the very space of their experimental set-up, and to which they cannot respond except by hypotheses or ad-hoc extensions[7]—in order to draw up a coherent picture. Such a picture will reveals not so much the imperfections of the theoretical apparatuses under examination, but rather the inappropriateness of the presuppositions (categories, principles...) of the empirical sciences. It will reveal also in contrast the notions and problematic dimensions under the light of which the phenomena examined must be approached and reflected.

What is immediately weighed in the balance is a causal versus expressivist conception of the stimuli and the reactions to them. In the first case, the stimulus is conceived in the form of an excitation which is at the source of a cascade of successively ascending and descending biochemical processes, and which conclude with the triggering of a motor action. In the other case, the stimulus is an object which has a meaning—a value—for the individual, precisely in that it concerns him or her and constitutes a motive for action for him or her.

In the second case, the vocabulary is the subjective vocabulary of intention, utility, value, purpose, meaning: terms of which Merleau-Ponty immediately wonders whether they should not be conceived as modes of "intrinsic determinations of the organism" (SoB: 10) and which presuppose "a new mode of comprehension" (SoB: 10). In any case, this vocabulary reflects the "immediate data of consciousness". For example, one could say that a luminous point attracts attention, attracts or draws the gaze: the event is therefore approached from within: it exists as perceived: « Common sense [recognizes] that one turns one's eyes "in order to see" » (SoB: 9).

However—this is the first case—from the point of view of an "objective study of behaviour" and a "scientific representation of the organism" (SoB: 8), as the stimulus in its "real" physical nature is not perceived: « it could not present itself as a goal toward which my behavior is directed. [the stimulus] can only be conceptualized as a cause which acts on my organism. » (SoB: 7).

In a « scientific » approach, the explanation of behaviour is based on the category of causality and the a priori of a "mutual exteriority of parts and processes" (*partes extra partes*, cf. supra): the stimulus as a more or less complex excitation (by constellation

[7] For example, the inhibition function introduced « [...] to justify the absence of an arbitrarily posited extension reflex. Here the idea is not introduced in order to render the fact itself intelligible but to mask a visible disagreement between theory and experience. » (SoB: 19-20).

or composition) and the reaction as a series of effects that it provokes can be broken down into "a multitude of partial processes, external to each other" engaged in a "longitudinal" chain and susceptible to "lateral" interactions. In short, the key concepts here are those of 'cause and effect' linking 'mutually external parts'.

From the organism as a biochemical complex to the organism as a living being in a world of values, we will move from a priori of cause and externality (of the parts) to a priori of expression and internal relations (between parts and from parts to the whole).

6.3.2 The Elaboration of Stimuli

In a "objective and scientific representation" (SoB: 9), the stimuli as causes are therefore material and defined independently of the organism they affect and the effects they provoke: the organism is passive: « it limits itself to executing what is prescribed for it by the place of the excitation and the nerve circuits which originate there » (SoB: 9). But this vision, even though it can be made more complex by various feedback loops, seems quite far from the reality of the interactions between the organism and its environment. For, as Merleau-Ponty observes after Weizsäcker, the individual, far from being a passive organism affected quite involuntarily by his environment, is best conceived as a "mobile keyboard" in search of its interactions: : « The organism cannot properly be compared to a keyboard on which the external stimuli would play and in which their proper form would be delineated for the simple reason that the organism contributes to the constitution of that form. [...] it is clear that each of my movements responds to a external stimulation; but it is also clear that these stimulations could not be received without the movements by which I expose my receptors to their influence. [...] The properties of the object and the intentions of the subject [...] are not only intermingled; they also constitute a new whole." [Thus] When the eye and the ear follow an animal in flight, it is impossible to say "which started first" in the exchange of stimuli and responses. » (SoB: 13). It follows that « [...] the form of the excitant is created by the organism itself, by its proper manner of offering itself to actions from the outside. » (SoB: 13). More precisely, the organism has the power to filter and reconform the various impressions it receives from its environment, by playing on the thresholds and chronaxies of its receptors or by the movement of its organs. In other words, it is the organism « [...] which chooses the stimuli in the physical world to which it will be sensitive, [and thus] "The environment (Umwelt) emerges from the world through the actualization or the being of the organism [...]" » (SoB: 13). More generally, « [...] the description of the known facts shows that the fate of an excitation is determined by its relation to the whole of the organic state and to the simultaneous or preceding excitations, and that the relations between the organism and its milieu are not relations of linear causality but of circular causality. » (SoB: 15).

This involvement of the organism in the constitution of its environment can at this stage still be understood as an operation of synthesis of the various impressions, certainly selected, but without any subjective inner dimension being involved. All of the above can in some way be "mechanised" or "implemented" in complex circuit diagrams and motor patterns: no awareness of a universe of objects with value, of an environment "worthy of interest", is required at this stage. This is in any case what Merleau-Ponty acknowledges with regard to the "transversal" relations between nervous circuits, each influencing the other: « [transverse relations] remain of the same type as the longitudinal relations of the classic conception: the organism plays no positive role in the elaboration of stimuli. » (SoB: 21).

However, experience forces us to go beyond a "real analysis of behaviour in isolatable fragments" as well as the assumption that these fragments would be causally related. Thus Merleau-Ponty reports that some physiologists have been led to introduce parameters of consciousness to account for the facts, in short « [...] to reintroduce quality into the language of science » (SoB: 15). For, as we observe, « when two excitants are in competition it is the painful excitant [...] which inhibits the other » (SoB: 15) : it is therefore the biological value of the stimulus that is decisive here.

The next step, in this demonstrative progression led by Merleau-Ponty, consists in untying the recognized meanings of "behaviours" from the material (biochemical) structures that supposedly instantiate them.Thus, he observes that one cannot correlate "fragments" of behaviour with well-defined neurobiological components: « [...] there was a tendency to assign, for each nerve element, a fragment of behavior which depended upon it: "verbal images" were localized; for each reflex movement a special device was sought. [But] More and more it was realized that the different nerve regions corresponded, not to real parts of behavior-to words, to such and such a reflex defined by its stimulus-but to certain types or to certain levels of activity: for example, to voluntary language as distinguished from automatic language, to flexion reflexes which, compared to extension reflexes, represent a finer adaptation, one of higher value. It is therefore a new kind of analysis, founded upon the biological meaning of behavior, which imposes itself [...] » (SoB: 20-21). This decorrelation of the biochemical and/or neurophysiological substrate organization of the behavioural structures manifests itself in an exemplary manner in the phenomena of irradiation: it can be observed that a sustained excitation does not diffuse continuously according to the physiological organization of the substrate, thus triggering processes which are increasingly distant according to such a topology, but gives rise to reactions which are characterized by the fact that they are provided with vital significance for the organism concerned: « With an increasing stimulation of the concha of the ear in a cat one obtains in turn: movements of the neck and of the front ipsilateral paw, movements of the back ipsilateral paw, contractions of the muscles of the tail and of the torso, movements of the contralateral back paw and movements of the front contralateral leg. Thus the pretended irradiation mixes symmetric and asymmetric reflexes, short or long, and does not invade the motor devices in the order in which they are anatomically placed. [So] "the fundamental forms of the movement of walking are what determine the character of the reflex much more than the spatial diffusion in the nerve substance." » (SoB: 25). And while the stimulus grows progressively, the organism does not respond with movements correlated to a continuous diffusion of excitement through pre-established circuits : « the excitation is elaborated in such a way that at each notable increase it is translated in the motor apparatuses by new movements and is distributed among them in such a way as to release a gesture endowed with biological meaning. » (SoB: 25).

The next step is decisive because it explicitly introduces the hypothesis, already sketched out but now made necessary, or at least finding its obligatory place in the picture thus drawn up, of the elaboration of stimuli.

Merleau-Ponty now discusses the theory of the composition of reflexes, which stipulates transverse mechanisms of inhibition: « [...] when a reaction is produced, all those which other stimuli could provoke at the same moment turn out to be inhibited; and when two antagonistic reflexes enter into competition in this way there is no compromise; only one of the two is achieved. » (SoB: 21). In the case of the processes which control the contraction of the flexor muscles, according to the previous hypothesis they should « [...] automatically provoke the inhibition of the extensor and vice versa » (SoB: 21). But this mechanistic vision proves to be too radical, because it is only valid in experimental configurations which separate the organism from its biological context, from its universe of actions, to install it in devices where it is no longer a question of vital behaviours or attitudes, in short, experimental set-ups which reduce the organism to mere biological matter.

For what is observed under natural circumstances is much more complex and nuanced than this hypothesis states. And it is again towards the behavioural meaning of the gesture, more or less "precise", "relaxed", "oriented", etc., that the relevant regulatory motif must be sought: the nervous distribution of the motor elements fulfils the value of the gesture rather than controlling it as a material and spatiotemporal process. In fact « [...] what appears to be a reciprocal inhibition is observed only if one employs electrical stimulation on muscles severed from their insertions. » (SoB: 21).

Thus, « Except for that of a strong movement, natural innervation does not follow this rigid law. As soon as it is a question of fine movements of the hand, or even of grasping movements, a simultaneous innervation of the antagonistic reflexes is observed, the distribution of which depends on the goal to be obtained and on the type of movement to be executed. Thus, it is not what happens at the level of the flexors which determines what happens at the level of the extensors, or inversely; but these two partial processes appear as aspects of a global phenomenon which must still be described. » (SoB: 21-22, emphasis added).

In the previous quotation, much of the challenge lies in the sentence « which must still be described », that requires a clarification of the process of « elaboration of stimuli ».

It is at this point that Merleau-Ponty appeals, without thematizing it, to a principle of simplicity. He takes note of what a material and causalist description of the functioning of the organism entails in terms of the cost of complexity: for in order to account for behaviour in all its contextual variations, it is necessary to constantly introduce new circuits of interaction or control so as to adjust the processes already hypothetically postulated. It is therefore advisable not to multiply the hypotheses as observations are made, and to do this, it is necessary to discover « [...] the central point of view by means of which all the facts could be coordinated » (SoB: 54).

In the particular case of antagonistic reflexes, it will be a question of « [defining] a conception of nerve functioning which renders intelligible at the same time and by the same principle the reciprocal exclusion of reflexes and the varied collaboration of the nerve circuits within each one of them. » (SoB: 22). And it is here that Merleau-Ponty introduced the hypothesis of the elaboration of stimuli: « If it were accepted that each reflex presupposes an *elaboration of stimuli* in which the whole nervous system is involved, one would understand rather well that it cannot "do two things at once" without the need of positing any special mechanism of inhibition. As to the regulated distribution of the motor excitations, it would find its explanation in precisely this same *elaboration of stimuli* which would be the proper function of the nervous system. » (SoB: 22, emphasis added).

This "elaboration of stimuli" will pertain to a holistic logic: the distribution of nerve impulses will no longer be conceived as a causal reaction to external stimuli, but as a regulation by a central system which filters, recomposes and redistributes the actions of the outside world in such a way as to trigger « [...] to release gesture[s] endowed with biological meaning. » (SoB: 25). It is « [...] a new type of order no longer founded on the permanence of certain circuits but created in each movement by the proper activity of the nervous system and according to the vital exigencies of the organism » (SoB: 25). This is the case, as we have seen, with irradiation, which provokes « [...] the vital movements of each animal instead of conforming to the anatomical distribution of the motor commands » (SoB: 228, note 48).

Following this view, the processes are not carried out according to pre-established circuits and their possible interactions, but « [...] depend upon the total state of the nervous system and on the active interventions which are necessary for the conservation of the organism. » (SoB: 26). Merleau-Ponty continues: « [but] How is this dependence of the parts with respect to the whole to be understood? » (SoB: 26) : how can we conceive of an "elaboration of stimuli" by the organism in such a way that their effectiveness is based on a biological meaning? In any case, we have moved from a passive conception of the organism to an active conception, where by means of "elaboration", the stimulus no longer influences as a physical agent but by the meaning it has for the organism: « Thus the excitation will never be the passive registering of an external action, but an élaboration of these influences which in fact submits them to the descriptive norms of the organism. » (SoB: 28). In other words: « The adequate stimulus cannot be defined in itself and independently of the organism; it is not a physical reality, it is a physiological or biological reality » (SoB: 31). It remains to elucidate the principle of elaboration of such a biological stimulus.

To achieve this, Merleau-Ponty will once again use the complexity argument. Precisely, in the discussion he opens on the ocular fixation reflex, he notes that the motor processes involved can vary with constant retinal excitation, and that to account for this within the framework of classical physiology, « An extremely complex shunting mechanism would be necessary [...] » (SoB: 34). Merleau-Ponty then continues: « Would it not be simpler to admit that the movement of fixation results, not from the addition of two series of excitations [external and proprioceptive], but from a total process in which the portion of retinal excitations and that of the proprioceptive stimulations are indiscernable ? » (SoB: 34). Continuing his demonstration, and considering certain reflex behaviours of the newborn, Merleau-Ponty then reports an overlapping of proprioceptive information and motor influxes. Thus, through proprioception, exteroceptivity and motor control are discovered in one piece. However, the unity of the sensorium and motorium thus recognized is not without resistance. It is easy to understand that mechanically speaking, this solution where « [...] the sensorium and motorium function as parts of a single organ. » (SoB: 36) is unthinkable. Indeed, how can it be conceived that flows constituted as such in their physico-chemical nature are reconfigured by the motor engagements that these flows would actually control? It is the therefore necessary to consider a type of design other than mechanistic, a type of design that is already required in fact by the unit, observed in the new-born baby, of proprioception and motor control. For from a mechanistic point of view it is not clear how an internal state, as a set of parameters describing the configuration of the body system, should trigger one motor action, and why this one rather than another. In order to allow this, we need a conception of proprioceptivity where the internal perception does not simply relate to a certain organic configuration but includes the awareness of an imbalance, of a kind of lack, which then contains the principle of its resolution, and which is therefore likely to trigger the movement towards a state where the said lack would be filled or towards a rebalanced situation.

At this moment, and as soon as the motor action is motivated by a more or less conscious appreciation of the value for the organism of the bodily (proprioception) or environmental (exteriorception) configuration in which it finds itself, we understand that sensorium, motorium and proprioception participate in the same logic. It will therefore be necessary to think of stimuli in a different order from that of materiality, that is to say according to an order in which they act not as a cause but as signs, and very precisely here as a signs of a motor meaning. We then see what is meant by the elaboration of stimuli: not a transformation processes with « constant ontological value », but the promotion of a material which is still non-existent for the organism at the rank of an authentic stimulus, that is to say at the rank of a percept invested with a biological meaning, in other words at the rank of an expression in the exact sense of the word.

It is then understandable why the stimulus can be paradoxically presented as "a response from the organism": « For the excitation itself is [...] not an effect imported from outside the organism; it is the first act of its proper functioning. » (SoB: 31). So, for example, the painful nature of an excitation, which is itself pointed out to the organism in this way, contains the motor idea of a strong protective reaction. In other words, « The notion of stimulus refers back to the original activity by which the organism takes in excitations which are locally and temporally dispersed over its receptors and gives a bodily existence to those beings of reason such as the rhythm, the figure, the relations of intensity and, in a word, the global form of local stimuli. » (SoB: 31).

All that remains to be done is to give a status to this physico-chemical "frémissement" that affects the receiver fields. In Phenomenology of Perception ([21], henceforth PhP), Merleau-Ponty will approach it in terms of "solicitation" and of "interest". In this preparatory work that is SoB, the stimulus in its physical nature is only an "opportunity" to perceive: a "circumstance" which is offered to the body and expects it to exist at its level of biological reality: « That which necessarily releases a certain reflex response is not a physico-chemical agent; it is a certain form of excitation of which the physico-chemical agent is the occasion rather than the cause. » (SoB: 31). Hence the equivocal nature of the notion of stimulus, "« [which] includes and confuses the physical event as it is in itself, on the one hand, and the situation as it is "for the organism," on the other, with only the latter being decisive in the reactions of the animal. » (SoB: 129).

In its widest scope, this principle of an "elaboration" of material excitations into percepts with expressive value, relates the fact of an intimate and harmonic connection of the organism to its behavioural milieu, as revealed in the so perfect twinning of body and environment. It is indeed because the organism has the power « [...] of modifying the physical world and of bringing about the appearance in the world of a milieu in its own image. » (SoB: 154) that it adjusts itself so perfectly to it. This is demonstrated in particular by the flexibility, assurance and adaptability of gestures to their environment. For if this is so, it is because the space in which the body circulates and deploys its actions is not a space configured outside of it, prior to its existence, in which it is immersed: it is a geometry of action that emanates from its motor power: « This space is bound up with the animal's own body as a part of its flesh. » (SoB: 30).

The result is a conditioning relationship between the organization and its environment, such that one (the environment) obliges the other, who in return confirms it. In the most elementary forms of the living, the contents of perception, in so far as they express latitudes of action, logics of taking, then determine the triggering: « Finally, [...] perception opens on a reality which solicits our action rather than on a truth, an object of knowledge » (SoB: 169). This obligatory conformation of actions to the perceived world is observable in certain pathological cases, at both bodily and verbal levels.

Discussing these questions in PhP, Merleau-Ponty then distinguishes "automatic" or "concrete" behaviours from "categorical" or "abstract" behaviours.

6.3.3 Concrete/Abstract

If we call "background" (of movement) the geometry of meanings that the body, as a vital power, institutes as its environment-the background of movement is not « [...] a stock of sensible qualities, but [...] a certain manner of articulating or of structuring the surroundings. » (PhP: 117)—and taking up what has been said about the body/world circularity, we understand that movement "adheres to its background" or, correlatively, that the background fully traces the lines of movement : « [...] the movement and its background are "moments of a single whole." » (PhP: 113), and the background of movement « [...] is not a representation associated or linked externally to the movement itself; it is immanent in the movement, it animates it and guides it along at each moment. » (PhP: 113). Such is the case at least for concrete movement, the "movement necessary for life" by which an ecological framework takes shape. For this type of movement, movement and background mirror one another, are contiguous to one another: It is "[from the affective situation of the whole that] the movement flows" (PhP: 107) and it is movement, as appropriation of the world, which institutes the characters of the situation where it unfolds.

But, continues Merleau-Ponty, a distinction must be made between "abstract" and "concrete" movements.

If bodily space is what circumscribes action, this only holds for concrete movements-in general, an object or a tool "are presented to the subject as poles of action; they define, through their combined value, a particular situation that remains open, that calls for a certain mode of resolution, a certain labor." (PhP: 108-109). Concrete movement is the direct and appropriate response to this solicitation for action, which "obtains the necessary movements from [the subject] just as [...] the customs of our milieu [...] obtains from us the words, attitudes, and tone that fits with them." (PhP: 109). Concrete movement, in a radicalized form induced by pathological states, rather than being a flexible and adaptable response to the solicitations of action, appears to be but a forced response, as if the environment imperiously commanded the subject to perform the required movements without any choice. In such cases, the affected person acts to satisfy the virtual lines that impose and guide his gestures: "[the patient] experiences movements as a result of the situation, [he and his] movements are, so to speak, merely a link in the unfolding of the whole." (PhP: 107).

And so, the gesture and its background are understood to form a whole: The movement and the situation become one, the gesture institutes an environment and a geometry of objects as signifying presences and these, in return, canalize the action of which they express the unfolding-thus, the affected person only succeeds in performing the "on command" (PhP: 107) : "[only] on condition of placing himself into the spirit of the actual situation." (PhP: 107).

But if the concrete movement only concerns, in its imperious form, the register of vital gestures, it also pertains to "habits": The "assimilated" gesture is indeed what responds to an environment perceived as the fitting and guiding receptacle for its accomplishment.

On the other hand, abstract movements are free from conditioning by any more or less assimilated situations. The abstract movement is "on command" and "[is] not directed towards any actual situation." (PhP: 105).

To accomplish an abstract movement is in a way "to possess my body independently of all urgent tasks, in order to make use of it in my imagination." (PhP: 115). In the pathological situations described by M.-P., these movements are very difficult to accomplish for some patients who may lack this aptitude of relaying a definite program in "abstract" form to their motor projects, be it on their own accord or in response to directives.

Dually, the abstract movement projects its fabric of meanings so as to establish the theater for the unfolding of a new situation. Being capable of abstract movements entails liberating oneself from the situation's conditioning through them. It means to cause things to recede and to introduce a plane of novel meanings between them and the acting subject: The motor project of abstract movement "aims at my forearm, my arm, my fingers, and it aims at them insofar as they are capable of breaking with their insertion in the given world and of sketching out around me a fictional situation"' (PhP: 114)—the abstract movement « [...] hollows out [within the busy world in which concrete movement unfolds], a zone of reflection and of subjectivity, it superimposes a virtual or human space over physical space. » (PhP: 114)—in other words, if "[the concrete movement] adheres to a given background, the [abstract movement] itself sets up its own background." (PhP: 114). With abstract movement, we "invert the natural relation between my body and the surroundings." (PhP: 115). Furthermore, with abstract movements, subjects "polarize the world, causing a thousand signs to appear there, as if by magic, that guide action, as signs in a museum guide the visitor" (PhP: 115); they are capable of "marking out borders and directions in the given world, of establishing lines of force, of arranging perspectives, of organizing the given world according to the projects of the moment, and of constructing upon the geographical surroundings a milieu of behavior and a system of significations that express, on the outside, the internal activity of the subject." (PhP: 115). Abstract movement therefore "a voluntary movement [which] takes place in a milieu, against a background determined by the movement itself" (PhP: 139)—that is to say that the movement projects its background.

Concrete and abstract movements are penetrated with two sorts of consciousnesses of the surrounding world: Concrete movement occurs in a world that is perceived and experienced as being built according to a guiding schema, whereas abstract movement operates within a world recognized as an "objective environment", a world of objects liberated from one's own body's power of action, instituted in themselves, and which, in return, require nothing from it.

Abstract movement, in its power to produce an environment of novel values, would thus present two concomitant facets: The residual side of a world devoid of its former "concrete" meanings, in sum, the "objective and impersonal milieu" of a transcendental consciousness, and a projected world of which the "objective milieu", passive and mute, constitutes a possible receptacle.

These observations also concern language facts.

Indeed, reinvesting in Gelb and Goldstein's work, Merleau-Ponty observes that certain linguistic troubles affect the capacity of subjects to employ words outside of their "concrete" context of usage. Thus, just as patients limited in their capacity for "spontaneous" movement can only move in some manners if the environment invites them to do so (the movement is then "concrete" because it "adheres to its background"), likewise, some patients find themselves incapable of speaking other words than those which constitute a verbal reaction to the situation: "The same word that remains available to the patient on the level of automatic language escapes him on the level of spontaneous language." (PhP: 180).

6.3.4 *Interior Relationships*

In the picture that we've sketched out above, the body and its world are effectively in a state of osmosis, in a fusional relationship of co-constitution: « [...] two correlatives must be substituted for these two terms [the body and the world] defined in isolation: the "milieu" and the "aptitude," which are like two poles of behavior and participate in the same structure. » (SoB: 161). And the key word that relates such a link of attunement is "circularity". Let us recall « [...] the relations between the organism and its milieu are not relations of linear causality but of circular causality. » (SoB: 15). Moreover, « [...] the organism itself measures the action of things upon it and itself delimits its milieu by a circular process [...] » (SoB: 148).

This character of circularity in the establishment of a body and its world is correlative to a relation of interiority from one to the other—therefore a relational modality specific to the expressive phenomenon (cf. supra), and which, as has been said, contravenes the order of material objectivity, namely the relation "partes extra partes": « Situation and reaction are linked internally by their common participation in a structure in which the mode of activity proper to the organism is expressed. » (SoB: 130).

And whether it is a question of bodily actions taken as such, or as turned towards a milieu, or as manifesting an intention of the subject, it is indeed each time an order of inner relations that is accomplished, in other words, an order of signification. For the gesture is like a melody, which is known to be present in each note, or like a sentence, of which every word accomplish the meaning: every moment of the gesture is inhabited by the intention of a body and the finality of an act. Thus, just as the act of speech bears its end and thus its totality from the first word, the movements « [...] that I execute with my body [...] anticipate directly their final position [...] » (PhP: 96-97).

In this holistic perspective, where the totality takes precedence over the parts and determines both their identities and their connections, the unity of the various parts is a "unity of meaning": For if the parts merge, it is because they share the idea (the concept) of the totality that they compose, in the sense that this idea institutes them and controls their synthesis (teleology) into an actual totality to the elaboration of which, therefore, they are discovered to be dedicated. Each part is thus internally connected to the totality that it accomplishes at its level and whose meaning it thus carries.

As regards the aim of the gesture, inasmuch as the gesture is aimed at and responds to a milieu and its solicitations, it is once again links of interiority that weave its execution: « Situation and reaction [...] are two moments of a circular process » (SoB: 130), or again, at the level of more elementary processes: « there is reciprocal action and internal connection among the afferent excitations on the one hand, the motor influxes on the other » (SoB: 47).

But there is more. The phenomena of the living also include in their significant dimension the expression of an interiority: As we have seen, the gesture means, on the one hand, by its holistic structure (« [...] the reactions of an organism are not edifices constructed from elementary movements, but gestures gifted with an internal unity. » (SoB: 130)), on the other hand, by its dialogue with an environment (« [...] to act upon [things] is to make an intention explode in the phenomenal field in a cycle of significative gestures, or to join to the things in which he lives the actions which they solicit [...] » (SoB: 189)), but also by its expressive rendering, i.e. by the exteriorization of an inner animation: « Our intentions find their natural clothing or their embodiment in movements and are expressed in them as the thing is expressed in its perspectival aspects. » (SoB: 188).

6.3.5 *Transition and Conjectures*

Having reached this stage, the questioning is twofold: For one, it bears upon the possibility of escaping the determinism of the body/world circularity, that is to say how to extricate oneself from the envelope of action that constitutes a "milieu". In other words again, and taking up Merleau-Ponty's terminological options, how to break relationships of interiority, so as to escape the "concrete" in order to achieve the creative freedom of the "abstract". Transposed to the semiotic plane, this line of questioning can be understood to bear upon the possibility of getting out of the expressive unity which, in the same way, determines and encloses the usage of signs when they have the value of expression, in such a manner as to attain access to the sphere of a free language creative of new meanings? It is clear here that an asymmetrical polarization of the sign, accompanied by the possibility of going beyond the regimes of legality proper to the order of signs, will be the key.

And secondly, what about this originary envelope of solicitations that the organism grasps? and How are these solicitations "elaborated" into perceptions with expressive value, i.e. elaborated into sensitive data "internally" linked to bodily meanings ? We will see that our problematic of heterogenesis is capable of providing the answer to each of these two lines of questioning in a single gesture.

For the moment, let us return to the theme of 'solicitations', and to the existential analysis of the act of perception.

6.3.6 *Solicitations*

Let's go back to the beginning. As we have seen, in the Merleau-Pontian perspective "all begins", to put it as such, with an interested and interrogative meeting between a bodily schema and an environment of solicitations, one which directs towards a crossed constitution of body and world, and having, from the onset, a value as co-expression. Thus, M.-P. emphasizes motor projects, the rhythms of existence, the solidary differentiation of sensible things and sensorial modalities, to posit the body as the central actor of an "expressive saga", inasmuch as it outlines through each of its gestures a world of signifying presences. A first modality of Merleau-Pontian being-in-the-world would therefore correspond to a desiring interrogation sparked by an environing halo within which meetings are sought. The local figure of such exploratory experience is the gesture, which is itself already in itself a response to a prior solicitation, an interrogation directed towards its source and an attempt at obtaining acknowledgement of it.

It is therefore necessarily the body as a carrier and performer of a certain life force and a hazy environment which "vaguely solicits", a sort of "poorly formulated question" (PhP: 222) : "Without the exploration of my gaze or my hand, and prior to my body synchronizing with it, the sensible is nothing but a vague solicitation" (PhP: 222)—with which I will attempt to syntonize and the effect of which will flourish into sensible qualities.

There is, thus, from the outset a set of diffuse solicitations; and the body is conceived as a muffled resource for action, with its life force steered towards surroundings, interrogating and attempts to meet or to appropriate what it encounters therein. But the appropriation by the organism of its interpellative environment requires in return a validation. This means that "In this exchange between the subject of sensation and the sensible, it cannot be said that one acts while the other suffers the action [...] » (PhP: 222). For if « My gaze subtends color, [if] the movement of my hand subtends the form of the object [...] » (PhP: 221), it remains that "My attitude is never sufficient to make me truly see blue or truly touch a hard surface" (PhP: 219). Very precisely: if the sensible in its latent state proposes to the organism "a sort of confused problem" (PhP: 219)., it is because it expects an appropriate solution: "I must find the attitude that will provide it with the means to become [some] determinate [quality]; I must find the response to a poorly formulated question. And yet, I only do this in response to its solicitation." (PhP: 219) Perception will primitively and fundamentally be this aptitude of receiving solicitations and, dually, of syntonizing with them so as to establish them within a world of objects and of qualities which are the expression, the living meaning, of this successful coordination: "the subject of sensation is a power that is born together with a certain existential milieu or that is synchronized with it." (PhP: 219).

In this "elaboration" of stimuli (understood here as clouds of solicitations), a double game is played, so to speak, without a discernible first move: a vital power promotes a purely interrogative scintillation to "bodily existence", i.e. installs such a scintillation on its own account as a perceptive phenomenon, a purely interrogative scintillation, but under the condition that the latter "finds itself in it": The sensible gives back to me what I had lent to it, but I received it from the sensible in the first place." (PhP: 219) To perceive is therefore to already be capable of being drawn into a certain fabric of solicitations, namely as it indistinctively resides in the laterality of our perceptual fields. Thus, to see an object is "either to have it in the margins of the visual field and to be able to focus on it, or actually to respond to this solicitation by focusing on it." (PhP: 69-70).

In its simple attentional aspect, the perceptual act thus consists in promoting to a higher level of determination a certain marginal and uncertain component of the field, for example through an emphasis that thematizes in function of the figure/ground articulation: "To pay attention is not merely to further clarify some preexisting givens;

rather, it is to realize in them a new articulation by taking them as figures." (PhP: 32). Thus, perception, in its attentional moment is a "passage from the indeterminate to the determinate." (PhP: 33).

But this marginal and uncertain portion of the perceptual field, its power of attraction, stems from its being full of announced presences, from its promise of a universe of things to be met-things that "are only pre-formed as horizons." (PhP: 32) In other words, "attention [...] is the active constitution of a new object that develops and thematizes what was until then only offered as an indeterminate horizon." (PhP: 33).

6.3.7 *Heterogenesis: From Sollicitation to Cosubstantiality*

Let us first return to the Hjelmslevian device, which we know places the semiotic function "above" forms and substances, and thus outside the field of semiolinguistic knowledge (cf. supra). If Hjelmslev proceeds in this way, he is certainly right to do so. For, as Hjelmslev will have seen, the semiotic function is not a phenomenon in the sense of empirical knowledge: the semiotic function cannot be apprehended in the manner of a substance whose form must be unveiled or whose laws governing its manifest functions must be highlighted: its intelligibility is of a different order.

Correlatively, the obstruction with which semiotic thought is confronted proceeds (cf. supra) from the fact that it views the planes of expression and content on an epistemic level where they come under the forms of empirical knowledge. Accordingly, and insofar as each of them is provided with its own substances, their unity becomes unthinkable.

It is therefore beneath all form and all matter, and therefore beneath the schemes of empirical rationality that are correlative to the a priori of form and substance, that the proper reason for the semiotic fact must be sought.

In any case, this is the problematic line to which MP invites us, in that he envisages the crossed constitution of a body and a world, both resulting from an interplay of interactions, where the body, initially posed as a deaf vital power, and responding to the uncertain solicitations of an environment which calls out to it, instructs it in return with its own rhythms, its specific behaviours, to then install in its exterior a world of sensitive qualities (the empirical diversity in the sense of classical epistemology). In this movement of co-constitution, sensible qualities are, by construction, intrinsically significant: the sensible is from the outset endowed with a meaning—namely, the meaning assigned to it by the bodily matrix that, as we've considered, institutes it. And the world in its native form is a world of expressions, in short, a world that already has value and meaning.

In order to advance in this direction, it will therefore be necessary to situate oneself on a problematic level prior to that of the concept of matter (in Hjelmslev's sense)—a concept which, as we know, makes it possible to inscribe in the gnoseological system this undifferentiated and therefore unknowable base which a relational framework will later inform.

Indeed, let us recall (cf. supra) that defined as a constellation of entities that are in themselves univocal, matter is configured from the outset as "homogeneous", in that all its elements share a common nature and qualities that, admittedly, leave them indistinguishable, but which dually bind them into an "amorphous" substratum where relationships can then be established in order to institute empirical objects.

To place oneself beneath matter is therefore doubly to place oneself at a distance from any formative modality and to position oneself in a place where the properties of homogeneity and continuity are not previously satisfied. In short, it means opting for an a priori of radical heterogeneity, which precedes the interplay of forms and substances, and therefore very precisely it means positioning oneself in a place where the possibility and production of substances is woven. All this is in close resonance with the views of MP, for whom it is a myriad of mutually irreducible, singular and unqualified solicitations, which are originarily offered to our vital behaviours—solicitations which therefore belong to a field of existence where determinations are not yet acquired, and to which even the most minimal form of homogeneity cannot be attributed without abuse. It is, for this reason, inaccurate to conceive of the elementary elements of matter (as an originary experiential source) in the former of simple undifferentiated units whose whole comprises a homogeneous mass.

Thus, whether it is a question of an originary installation of a signifying world in relation to a proper body (M.-P.) or of the primordial fact of an interpenetration of the planes of expression and content (Hjelmslev), it is each time beneath all components or constituted dimensions that the elements of an explanation must be sought, and therefore below the hypothesis of primary units that constitute "homogenous" matter and that serve as a potential support for a more cohesive form (Hjelmlev), or below stable and determined sensitive qualities (M.-P.).

The examination of the fundamental forms of semiotics must therefore begin at this level, recognized by M.-P., where a multitude of local tensions, mutually irreducible in the sense that they do not weave common matter, constitute the originary milieu, and within which, by a sort of tightening towards homogeneity and continuity, the constitution of current flows or aggregates can be envisaged and studied.

To give ourselves the means to do so, we can consider a multitude of differential operators, mutually irreducible, which, in that they are each defined locally and concentrate, in their intensive sense, universes of possible forms, relate the essence of the "originary diversity of local tensions" previously considered.

This being the case, mathematical tools presented in previous chapters are available to qualitatively explain how this radically heterogeneous diversity can be overcome in favour of a kind of "weaving": where these tensions, initially simply intensive and mutually alien, come into contact to constitute the actual extension of an empirical substratum.

More precisely, we will first build a "quotient" space that assembles and unifies the tangent spaces attached to each operator. Then (Rothschild and Stein theorem—[33]) we show that any curve that can be integrated into this quotient space attests to a solution common to the operators attached to the initial and terminal points of this curve. This result is remarkable in that, on the sole basis of the tangent spaces, its conclusion is valid, through their general solutions, for the differential forms of all complexities (parameters, compositions and degrees). But, as far as we are concerned, what this theorem essentially expresses is that the existence of an actual form (the integrated curve), and as this curve attests to the possibility of a solution common to various operators, is therefore correlative to a connection of the local domains on which these operators find themselves defined. In other words, the data of a current form, as it is integrated in the tangent space, is like the echo within an extensive plane of a contact, by way of partial overlapping, of the local domains of the intensive operators. The existence of a current extensive form thus is the expression of a certain resolution in the intensive plane of the originary heterogeneity carried by the differential operators.

This being the case, the existential scenario which is set up on these supports is the following: at the beginning, therefore, and even below the originary polarity between a vital power, in search of a body, and an environment of solicitations waiting to make a world, we will envisage a "pulverence" of local tensions, rendered through differential operators, a sort of intensive value nodes where the passages to an extensive actuality are woven, and which, in a way, seek each other out in order to pass into existence. Such an extensive existence, which, as we have just seen, is correlative to a "rough" continuity of the operators, and through the introduction of a temporal parameter, can be seen as the installation of a generalized vital flow, preceding any interiority or exteriority, and as such anterior to any body/world distinction. Now, and this point is crucial, this flow, which is thus a response to the various intensive local tensions aspiring to an integrated actuality, is configured through a harmonic analysis, on the basis of its eigenvectors which, deliver its resonance dimensions (or main dimensions). In other words, when grasped at its native stage, it appears that the empirical space of a materiality 'in general' is the correlative of a resonance process whose ranges (the plateau in Deleuze's sense) then constitute intrinsically paired substances, that is to say, expressed in terms previously introduced and discussed, substances maintaining relations of interiority (vs. relations *partes extra partes*—cf. supra), or again: relations of cosubstantiality. In this way, the heterogenetic set-up, which, as the integration of a diversity of intensive operators, formally expresses the promotion of a diversity of solicitation, especially sensory, to actuality, the heterogenetic set-up, therefore, and in that it is problematized at a level underlying the constituted substances, makes it possible to give an account of the elaboration of a co-substantiality. From this heterogenetic perspective, the planes of expression and content are discovered to be in a relationship of interdependence, in that they constitute the dimensions of resonance which structure the realization of a flow or an actual materiality.

However, the dimensions (or sub-substances) thus identified, carried by the main vectors, are at this stage amorphous. It will thus be necessary to conceive further, in addition—and the morphodynamical apparatus lends itself fully to this—to conceive of a principle of mutual structuring of these dimensions, so that on each one of them, qualified components are configured by way of their relations with others. Having accomplished this, we will have rediscovered, or better, we will have reconstructed, the famous Saussurean schema ([3]: 156) in which A and B are the substances of expression and content; by way of our reconstruction, however, these would now be conceived as the actual resonance ranges of a diversity of solicitations, in which the vertical cuts structuring these ranges differentially and dually are the result of morphogenetic processes, which a morphodynamic approach to the Saussurean sign fully accounts for—cf. [31] & (supra).

6.4 A Convergent Argument

6.4.1 Overview

Considered as such, and as recalled above, the expressive fact appears paradoxical, aporetic and unfit for any conceptualization. However, its rationality is not null, and it becomes accessible "from below"—that is, when we approach it not as a stabilized phenomenon, delivered to a work of conceptualization, but in its logic and with respect to the principle of its formation, namely in terms of "elaboration" as developed by Merleau-Ponty.

The Merleau-Pontian scenario, let us recall, is that of a certain vital power, provided with an inner norm, which finds itself worried or questioned by a multitude of uncertain solicitations: tensions, fluences or points, which are so many "poorly formulated questions", and which make a claim for a "bodily existence" with regard to the organism that they interpellate. This "bodily existence" will be attributed to them through an appropriate commitment of the organism towards them, the organism thus instituting them in their form of empirical presence, namely as qualitatively defined data of perception.

We thus witness the installation of dimensions of experience: forms of intuition, sensitive qualities, which are not abstract or inert qualifications, reducible to a geometry or to measurements, but which are the tangible correlates of a certain way of going towards... of seizing... (here we overlap with the notion of motif—cf. [2]) as a response to a cloud of solicitations: « Blue is what solicits a certain way of looking from me, it is what allows itself to be palpated by a specific movement of my gaze. It is a [...] certain atmosphere offered to the power of my eyes and of my entire body. » (PhP: 218).

Now, for our purposes, we shall retain that phenomenal qualities are the fruit of a successful, almost osmotic, encounter between a living organism, a pure vital power, and a latent environment which interests it and to which it adjusts an appropriate response. That is to say that sensible quality is a fundamentally expressive value: its "sensory" reality is nothing other than an empirical installation of the vital forces brought into play to respond to a latent environment in search of a "bodily existence" as a "milieu": « [...] the sensible does not merely have a *motor and vital signification*, but is rather nothing other than a certain manner of being in the world that is proposed to us [...], that our body takes up and adopts if it is capable [...] » (PhP: 219, emphasis added).

6.4.2 From Expression to Speech: The Problem

In the sensory fact there is thus a reciprocal incorporation of a motor signification and a sensible quality, and this "incorporation" is no longer a matter of metaphor, but can be thought of according to the dynamic modalities exposed above, which the heterogeneous approach accounts for.

By generalization, and without going into unnecessary details, we can then conceive of the expressive fact in the mode of the simultaneous installation, on the plane of empirical existence and in an undivided form, of a certain content (no longer necessarily motor) and a certain expression (in the sense of Hjelmslev). And it is with an expressive material thus configured that we will now approach the "problem of primordial speech".

The question of primodial speech' is addressed by Merleau-Ponty from at least two analysis angles, the first in PhP, through a problematic of gesture, and the second in Signs, in terms of differentiality.

6.4.3 From Speech as Gesture to the First Speech

It is in PhP page 208 that M.-P. introduces the conception of a word as a gesture. After confirming the "expressive" character of the word, i.e.: « [...] then, the sense of words must ultimately be induced by the words themselves », (PhP: 184), he continues: « or more precisely their conceptual signification must be formed by drawing from a gestural signification, which itself is immanent in speech» (PhP: 184). Fundamentally, according to Merleau-Ponty, if speech can claim to have a gestural nature, it is because it principally pertains to an existential order. To speak is not simply to assemble words into signifying combinations, but to act in the world instituted by a language, as one's body acts within its environment. And to know how to speak is not to possess a system of rules and conventions, it is to know one's possibilities for action within the universe of experience of language: "[...] I begin to understand the sense of words by their place in a context of action and by participating in everyday life, [so too] I begin to understand a philosophy by slipping into this thought's particular manner of existing." (PhP: 184-185).

Moreover, this verbal gesturality, like the gesturality of one's own body, generates its meaning-from which conceptual significations may therefore be retrieved. Thus, "we are clearly led to recognize a gestural or existential signification of speech, as we said above." (PhP: 199). In short, "speech is a gesture, and its signification is a world" (PhP: 190). But how? In what respect is speech to be assimilated to gesture? We will first observe that bodily movement and the act of speech both present a holistic and finalized character. Regarding speech, it is established that the utterance is not a summative succession of words, but indeed an integrated totality accomplishing a certain intent to signify. The act of speech carries its own end and therefore its totality from the moment the first word is uttered, likewise for bodily movement: "[T]he originality of movements that I execute with my body: my movements anticipate directly their final position [...] I do not find [my body] at one objective point in space [like an object] in order to lead it to another, [...] I have no need of directing it toward the goal of the movement, in a sense it touches the goal from the very beginning and it throws itself toward it." (PhP: 96-97).

But it is from the angle of their practice that the parallel between gestures and speech is the most flagrant. Indeed, similarly to when the empirical world has and delivers things following a geometry and a play of qualifications which express their immediate relations to a certain capacity of action, likewise language delivers a world of words and of constructions as they "constitute a certain field of action held around me." (PhP: 186).

To speak therefore amounts to moving through speech within a world of words: "I relate to the word just as my hand reaches for the place on my body being stung. The word has a certain place in my linguistic world [...]" (PhP: 186). And likewise that the body knows its world on the mode of a "power to do", speech knows words on the mode of a "power to say", which is therefore "power" by virtue of words: "[K]nowing a word or a language [langue] does not consist in having available some preestablished neural arrangements [or some verbal representations] [...] the words that I know [...] are behind me, like the objects behind my back or like the horizon of the village surrounding my house; I reckon with them or I count upon them, but I have no 'verbal image' of them" (PhP: 186), or: "Likewise [for movement], I have no need of representing to myself the word in order to know it and to pronounce it." (PhP: 186).

The gestural nature of speech then appears clearly. In the same manner as bodily movement installs a sensible world with respect to a subject which invests it, language "is the subject's taking up of a position in the world of his significations." (PhP: 199). And it is not a matter here of metaphors: The principle of constitutive interactions which simultaneously install a subject and his or her world just as much concerns the gesture of speech with respect to the world of significations. "The term 'world' is here not just a manner of speaking: it means that 'mental' or cultural life borrows its structures from natural life and that the thinking subject must be grounded upon the embodied subject." (PhP: 199). Furthermore, "the phonetic gesture produces a certain structuring of experience [...] just as a behavior of my body invests-for me and for others-the objects that surround me with a certain signification." (PhP: 199).

It is at this stage that certain difficulties arise.

For to understand a gesture, to grasp its immanent meaning, is therefore to seize for oneself, so far as this may be possible, the internal dynamics of this gesture that accomplishes a particular engagement of the body and that installs at its horizon a meaning which then becomes accessible: "The gesture I witness sketches out the first signs of an intentional object. This object becomes present and is fully understood when the powers of my body adjust to it and fit over it" (PhP: 191), or: "The sense of the gesture thus 'understood' is not behind the gesture, it merges with the structure of the world that the gesture sketches out and that I take up for myself." (PhP: 192). Likewise for the verbal gesture, as we have seen that through it the listener takes on and reanimates the significatory aim of his or her interlocutor.

But for this to be possible, it is necessary for the interlocutors to share the resources which inform a shared universe of existence. But unlike the bodily gesture, however, which is performed in a shared environment of (« [...] the gesture is limited to indicating a certain relation between man and the perceptible world, [...] this world is given to the spectator through natural perception, and [...] the intentional object is hence offered to the observer at the same time as the gesture itself" (PhP: 192)), the verbal gesture "[...] intends a mental landscape that is not straightaway given to everyone, and it is precisely its function to communicate this landscape" (PhP: 192). Thus, If speech is a gesture, this gesture is executed within the world which is its own: not a world of sensible objects and qualities, but a world of signs and meanings. And this world "[that] nature does not provide, [...] culture here offers [it]" (PhP: 192), such a world being a space for symbolization, a landscape of values and a framework of usages, which original meaning-intentions will invest to their own ends. This shared space in which verbal gestures accomplish themselves is therefore that of spoken (also called sedimented) language: It is "[the] available significations, namely, previous acts of expression, [which] establish a common world [...] to which current and new speech refers, just as the gesture refers to the sensible world" (PhP: 192)-and moreover, "the sense of the speech is nothing other than the manner in which it handles this linguistic world, or in which it modulates upon this keyboard of acquired significations" (PhP: 192)."

But a question then arises: that of "primordial speech". Because if the verbal gesture takes place within a world of significations which constitute the mental landscape of a community of speakers, it is necessary to explain how, starting from nothing, language was able to progressively install such a world. "For the miracle to happen, the phonetic gesticulation must make use of an alphabet of already acquired significations, and the verbal gesture must be performed in a certain panorama that is shared by the interlocutors, just as the comprehension of other gestures presupposes a perceived world shared by everyone in which the sense of the gesture unfolds and is displayed." (PhP: 200). We should insist: "[Since] speech is a genuine gesture and, [as] just like all gestures, speech too contains its own sense" (PhP: 189), it is indeed necessary for this gesture, in the manner of the bodily gesture, to have an environment which will instruct new meanings by virtue of its own power. Now, this environment, we have seen, is sedimented language, resulting from the accretions of speech. A "primordial speech" is therefore necessary in order to install a first world of significations.

The idea of a verbal gesticulation emanating meaning in an equal measure as one's own body would lead to an impasse. There would, however, be some advantage in pursuing its examination, particularly to find in the articulatory forms of the verbal gesture a new formula of its immanent sense.

In essence, according to M.-P., verbal gesticulation proceeds on the one hand by way of interlacings and overlappings, and on the other hand, by way of convergence and condensation. To speak is to put into sequence, to superimpose and to progressively integrate a series of elementary verbal gestures which in fine install at their fore, as the focal point of the tensions which animate them, a certain signification: "The clarity of language is not behind it in a universal grammar we may carry upon our person; it is before language, in what the infinitesimal gestures of any [...] vocal inflection reveals to the horizon as their meaning." ([22]: 28). In the exercise of authentic speech, it is therefore not a matter of serving again verbally encoded meanings, but of using words in a manner such as "The cross references multiply [and] more and more arrows point in the direction of a thought I have never encountered before." ([22]: 12). In short, the meanings of speech are like "ideas in the Kantian sense[:] the poles of a certain number of convergent acts of expression which magnetize discourse without being in the strict sense given for their own account." ([24] : 89).

A moment must therefore arise in which this coherent accumulation of punctual gestures, which are so many semantic adumbrations, ends up being crystallized and causing the rise of an object of meaning in consciousness: "Once a certain point in discourse has been passed, [the sketches (Abschattungen)] suddenly contract into a single signification. And then we feel that something has been said." (ibid.: 91). We can then say that thought will have been expressed "when the converging words intending it are numerous and eloquent enough to designate it unequivocally." (ibid.: 91). Language is therefore an "oriented system." (ibid.: 88). This must be understood not as a system which carries internal dynamics which would make it tend towards certain of its possible states, but a system which tends to produce an "exterior" where it installs the subject as if in a new dimension of experience: "A language is [...] a methodical means of [...] constructing a linguistic universe of which we later say-once it is precise enough to crystallize a significative intention and to have it reborn in another-that it expresses a world of thought, as it gives it its existence in the world." ([22]: 31).

From this, it is clear that meaning is "induced" by gesture: meaning is not itself present within gesture. This amounts to saying that the constituents of verbal gesture do not contribute to the sense of a discourse because all together they produce a convergence towards a certain meaning. The speech gesture is to be approached on a plane of functioning where its components are disjointed from meaning: A language—French, for example—is not a sum of words, "[it] is the configuration that all these words and phrases draw according to their use in the French language. This would be strikingly apparent if we did not yet know the words' meaning and were limited [...] to repeating their coming and going, their recurrence, the way they associate with one another, evoke or repel one another, and together make up a melody with a definite style." ([22]: 32).

Here we find again the problem of the first speech. For the principle of primordial speech is indeed that of a power to signify by means of words prior to any conventional association between form and meaning, such as those which are registered by sedimented language (as an action landscape of speech), therefore of a power to signify by means of words before their acquisition of meaning, and even before they are polarized into signifiers and signifieds.

But if the problems are thus better posed and better specified, they are not however resolved. Indeed, at this stage, we can always do without an internal significatory resource: Regularity, overlaps, etc., are not sufficient in themselves to induce a focal point which is external to the system. All such plays of repetitions and of systematicity suppose, in order to produce an effect of convergence, an internal animation remaining to be elucidated. In all likelihood, the cumulated effects of the intersections and overlaps "suggest even more that the whole process obeys an internal order, the power of revealing [...] what [is] in mind." ([22]: 33).

But it does not suffice to suggest that "all of this" is animated from within, because the suggestion does not give access to the principle of animation which alone installs a world of meanings. At most, from such a "suggestion", we may derive a system of semantic representations according to the relational distributions observed and following a modelizing approach. There is therefore an insurmountable gap between a convergence of systematic overlaps, which never exceeds its own order, and the convergence towards an exteriority, as an installation of a new dimension of experience.

For the time being, we will leave the question as it stands. For in order to move forward in its treatment, we need additional elements. But before completing this paragraph, it will not be useless to recapitulate the problematic position reached here in clear terms: if the first word is installed at the point of a series of non-semiotically formed verbal "impulses", very precisely as an intentional tracing configured by a set of elementary verbal gestures, then two things need to be clarified: On the one hand, the mode of relationship by which the series of elementary verbal pieces is coordinated into a system that determines a specific object of meaning, and on the other hand, the inner resource from which verbal gestuality draws and which, through the system of elementary actions that it performs, is thus installed as an intentional object of meaning.But we must beware of the misunderstanding of such a formulation. For the two components of the verbal gesture retained here are not "orthogonal" components: that is, independent and complementary in the elaboration of a speech object, and, in particular, do not position themselves mutually as form and substance. The fact remains that it is in the interplay of these two dimensions, thus proceeding from a common principle, that the linguistic gesture elaborates its power to « [...] secrete through its internal organization a certain originary sense upon which the significations will be outlined. » ([22]: 31). To approach this, we will take up the question of the first word from another angle, the one that Merleau-Ponty exposes in Signs 1964 [24].

6.4.4 The Differential Foundation of the First Speech

The thesis that Merleau-Ponty develops in Signs, and which accompanies his diacritical conception of perception and meaning (cf. in particular [20]), is that of a differential foundation of the semiolinguistic faculty: the child emerges from babbling when he accedes to a differential awareness of the motor and sensory events, and more fundamentally the expressive events, of the world with which he develops his commerce.

If Merleau-Ponty, after Saussure, recognizes the fundamentally differential character of the semiolinguistic fact (primacy of difference over the term), it is not exactly with the same gnoseological perspective as that of the Swiss linguist. For Saussure, the regime of difference is a category of the structural episteme and, in this sense, governs the constitution of semiolinguistic objectivities. If Saussure's intention, through the principle of negative identity, is primarily to access an objective recognition of linguistic factualities, Merleau-Ponty takes up difference essentially from the point of view of its holistic scope. Of course, this idea of the primacy of the system over the parts is also clearly defended by Saussure, but for Merleau-Ponty it is above all a question of drawing consequences from it. For what Merleau-Ponty insists on is that the consciousness of difference is coupled with to a consciousness of totality, and that the acquisition of one opens up a sphere that goes beyond its mere fact.

After recalling that, according to Saussure, language is made up of pure differences, i.e. "differences without terms" ([24]: 39), or conversely that « the terms of language are engendered only by the differences which appear among them » (ibid.: 39), and in order to respond to the paradox of learning (by which one goes from the parts to the whole, whereas it is the whole that is primary), Merleau-Ponty observes that when they are elaborated as differences, the parts immediately refer to the whole. For in a differential system, the terms are like sides of a shared border and as such are mutually adjoining, and hence share « [...] a unity of coexistence, like that of the sections of an arch which shoulder one another » (ibid.: 39). Therefore, since the idea of the system is induced by the differential nature of the parts, we understand why "In a unified whole of this kind, the [...] parts of a language have an immediate value as a whole [...]" (ibid.: 39-40).

Going a step further, for Merleau-Ponty it is in the phonemes themselves, defined according to Saussure as oppositional and relative entities, that the possibility of language resides:« [...] because the language in its entirety [...] is anticipated by the child in the first phonemic oppositions » (ibid.: 40). For it is the phonemes, « [...] which do not for their part have any assignable meaning, [that have the] function to make possible the discrimination of signs in the strict sense » (ibid.: 40). Thus, the phonematic fact includes, in addition to the idea of a systemic totality, the suggestion of an opening onto a signified, simply because « [...] the meaning arising at the edge of signs [is correlative on a new dimension] of the immanence of the whole in the parts [...] » (ibid.:). Merleau-Ponty here revives the problematic Saussurean thesis of the coincidence of "vertical" (between signifier and signified) and horizontal (between signs) relations. Indeed, still according to Merleau-Ponty, with phonemes « [...] the child seems [therefore] to have 'caught' the principle of a mutual differentiation of signs and at the same time to have acquired the meaning of the sign » (ibid.: 40). In other words, « [...] it is the lateral relation of one sign to another which makes each of them significant, so that meaning appears only at the intersection of and as it were in the interval between words. » (ibid.: 42).

This short presentation calls for two clarifications. Firstly, concerning the meaning of the "gap" that oscillates between (formal) differentiality and diacriticity, and secondly, concerning the intuition of the holistic engagement of a differential relationship.

With respect to the first point, it should be specified that Merleau-Pontian diacriticity does not cover Saussurean differentiality. In fact, the relationship of difference that Saussure sees between phonemes or between signifieds is a relationship that operates in substances, respectively, (the substance) of expression and (the substance) of content. We are therefore dealing with the problem of a form that instructs a homogeneous substrate, precisely in that it administers a categorization by means of the emergence of a system of differentiating thresholds; in other words, a form that, even though it is differential, proceeds to syntheses. The diacritical, for its part, is to be understood intuitively and in a first stage as a "distance with respect to a certain level", the grasping of a salience against a certain background. But it is, so to speak, an "existential grasping of salience"-in that this general structure of perception/expression articulates in its own order a background of matter which is indecisive and soliciting, an originary environment no longer posited and thought as such as being anterior to its grasp by the vital power of a bodily schema, but rather, thought as relative to the "promotion" of a thing thematically perceived or expressed and therefore taking position as a figure. The notion of background, a gestaltist legacy, is thus, within this new device, radically reworked and shifted. Beyond and beneath the sensitive surface of the field, it opens onto a very broad notion of a background of solicitations, motivations, horizons, and of praxis...

The second point to be discussed briefly is that of the holistic meaning of differential relations. Here, Merleau-Ponty's intuition must be acknowledged as exceptionally accurate. For in truth, nothing allows us to move a priori from the idea of a relation, even a differential one, to that of a systemic totality. No doubt, every relational fact presupposes a device that subsumes it, but this does not establish the holistic nature of the device. Within the framework of axiomatic-formal theories, for example, and given a relation R defined by certain axiom schemes, it is possible to conceive of an indefinite number of systems, without limit of complexity and without constraints of internal organization, where this relation would be taken in charge. In other words, there is no a priori link of necessity between the relational fact and the system to which it belongs. But as we are dealing with differential relations, of a topological and dynamic nature, this is not the case. For what the morphodynamic approach (Thom, Petitot) teaches us is the very high degree of constraint (induced by the condition of structural stability) that weighs upon the complexity and composition of morphologies. Thus the differential fact is not of the kind of connection that can be arbitrarily assembled at will, but proceeds from the requirements of a totality that subsumes and induces it (by way of stabilization of an originary singularity). In this case, therefore, there is a relation of necessity between the totality and the parts, which validates the Merleau-Pontian intuition.

Having said this, and returning to what concerns us, we will now see how and in what way the differential hypothesis of the first speech intersects, completes and resolves the question as formulated at the end of the previous paragraph—a question that, in Signs, Merleau-Ponty expresses as follows: how « The untiring way in which the train of words crosses and recrosses itself, and the emergence one unimpeachable day of a certain phonemic scale [...] finally sways the child over to the side of those who speak » ([24]: 40-41), or again: how « language [...] invents a series of gestures, which between them present differences clear enough for the conduct of language, to the degree that it repeats itself, recovers and affirms itself, and purveys to us the palpable flow and contours of a universe of meaning. » ([22]: 32).

6.4.5 *From Expression to Speech: Outline of a Heterogenetic Solution*

If we take up the previous quotation, we find the two ingredients of the problem discussed. First, there is « The untiring way in which the train of words crosses and recrosses itself » ([24]: 40) (or this "series of gestures" that repeat, overlap and confirm each other), which raises the question of the type of relationships at stake in these contrasts and matching. Then there is the thesis of the "swaying over to the side of those who speak", i.e. the elaboration of a system of signs and concomitantly the opening of verbal gestuality onto an "irrecusable" intentional object of signification.

Concerning this verbal activity in its originary "antesemiotic" form—that is to say, this series of preverbal oral-phonatory gestures that contribute to the establishment of a meaning—and situating ourselves in the Merleau-Pontian perspective previously reported, we now have this capital piece of information at our disposal: it is a question of phonemes, which are differential entities. The type of relations at work in the originary verbal gestuality are therefore relations of difference.

But, if the morphodynamic perspective that theorizes differential processes does indeed account for the installation of a system of phonemes, it is totally defective when it comes to approaching, beyond the phoneme as constituted, the discriminating power of the said phonemes at the level of signs ; this is the case because this discriminating power opens onto universes of meaning, namely this remarkable « [...] lateral liaison of sign to sign as the foundation of an ultimate relation of sign to meaning » ([24]: 40). For it is easy to see that the phoneme, approached in morphodynamic terms, thus as a differential value instituted within a substance of expression, is intrinsically devoid of any extension into universes of meaning.

Consequently, if we choose to retain the formula of a differential process at the source of the first speech, it will be appropriate to have it operate on a material "prior" to the phoneme as a structurally determined object, but which nevertheless remains linked to it. Moreover, this "prior" material will be expected to contain that tension which, triggered between the signs, and configured by their relationships, installs beyond themselves an object of signification.

To achieve this, one solution consists in situating oneself at the originary level of expressivity: to consider the phoneme as an expressive fact, prior therefore to its relational qualification which institutes its empirical objectivity (precisely as the differential identity of a substance of expression), and to reconsider the mode of differentiality at this level of reality. And since the expressive fact incorporates the orders of the sensible and the intelligible, and thus conceals the material from which to draw substances of content and expression, this formula provides an answer to the problem of the installation of a meaning with respect to the series of elementary verbal gestures.

In truth, in this operative picture, if it is a question of contrasting the "expressive morsels" [the word is Visetti's] which are linked, distinguished and overlapped, so as to draw "in dotted line" and for the consciousness of the speakers an object of meaning, it is not in the mode of a formal difference. This would be a contradiction in terms, since this kind of differentiation, which determines an object, is only operative in relation to substances. And, as we have seen, the expressive fact cannot be approached in these terms.

The scenario that takes shape at the crossroads of the problematics of gesture and differentiality is then more or less as follows: first of all, there are expressive events, sensory materials or motor actions that are intrinsically significant. There is also, it must be assumed, an intention to signify that seeks itself to take shape, and requires the surpassing of what it has at its disposal on the merely expressive level. Then begins, using the material available, namely some sort of "expressive morsels", clumsy attempts at adjustment, correction... "tireless overlaps", insistences and denials... all of which tend to inaugurate a necessary but as yet non-existent signification.

It is thus a question of going beyond what the available forms of expressivity deliver and allow. But these interplays of confrontation and contrast can only succeed at the cost of dislocating the expressive material mobilized. Firstly, because the intention is precisely to go beyond them, but above all because in the expressive fact the signified content, vital or motor in its primary nature, and the material that embodies it are perfectly inseparable, precisely on the grounds of "inner relations". The modality of deviation, confrontation, insistence, etc., which underlies these attempts must therefore be inscribed in a scheme where, in fine, signifier and signified are still dually constituted, but on separate planes upon which differential operations freed from the expressive straitjacket are then possible. It is at this point, in order to achieve this, that the principle of "elaboration" of expressive phenomena must be called upon (cf. supra). For the expressive fact, in order to accomplish the confrontational tension in which it is now engaged, a tension impossible to satisfy at the level of organization where it is posited, the expressive fact, therefore, has no other way, so to speak, than to undo its own state—acquired, let us recall, and in its originary formula, as the "elaboration" of a halo of solicitations delivered to a motor or vital power that responds to them—in order to recompose itself according to modalities in which relations of differences can be established.

It will thus be a question, in a certain way, of returning to the stage prior to the elaboration of an expression, to the stage where motor or vital animations are in front of a soliciting diversity, and before any syntony. At this point, the confrontational tension that animates verbal gestuality will be deferred below the expressive fact and possibly accomplished in the mode of a differentiation of motor animations, correlated to differentiations of the soliciting diversity—which supposes the setting

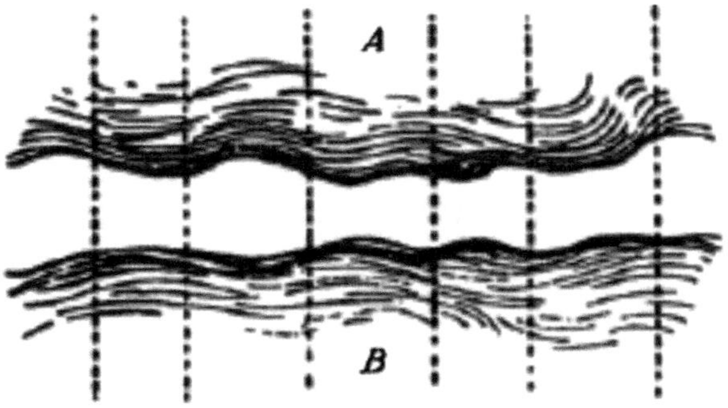

Fig. 6.1 Saussurean scheme of jointly instantiated subdivisions on amorphous masses of sounds and ideas

up of a new modality of "elaboration" on both sides: namely as substances of content and expression respectively.

In the end, the expressive gap is thus commuted into differences, at the same time as the incorporation of meaning into the sensible is unravelled into two correlated substances. We thus move from a scheme in which expressive data, whose sensoriality is in itself signification, manage to accomplish their confrontation at the price of a disintegration followed by a re-elaboration into substances of content and expression correlatively resoldered by a (differential) principle of common formation—a principle of common formation of which the Saussurean scheme of the two amorphous masses ([3]: 156) adequately accounts (see Fig. 6.1).

If we were to illustrate this transfiguration of the expressive fact into a semiotic fact, we could propose the following diagram (Fig. 6.2) (the myriad of solicitations is represented by asterisks and the vital tensions by wavy arrows): The impossible relationship of confrontation (double red arrow) between the two expressions (1 and 2) is realized through the form of correlations of differences (vertical red lines) in substances (of expression and content—SoE and SoC) instituted by taking from each expressive data (1 and 2) and placing in continuity, on the one hand, the vital powers, and on the other, the diversity of solicitations that respond to them. If one agrees with such an explanatory scheme, it is necessary to go beyond its main principles to provide the determination of its various strata and functional stages.

It is at this point that the heterogeneous device intervenes effectively. For, as we have seen, the heterogenetic approach accounts for the emergence of substances. Heterogenesis also allows us to conceptualize the reconstitution of expressive elements which, in the search for mutual confrontation, return to the state of "soliciting clouds" (rendered by a diversity of differential operators) in order to recompose themselves in a global expressive structure where "planes" of expression and content (corresponding to the dimensions of resonance of the integrated operators) are paired and possibly instantiate differential relations instituting phonemes and signifieds on both sides. It is this last structural configuration that a morphodynamics of the sign will precisely restore (cf. supra).

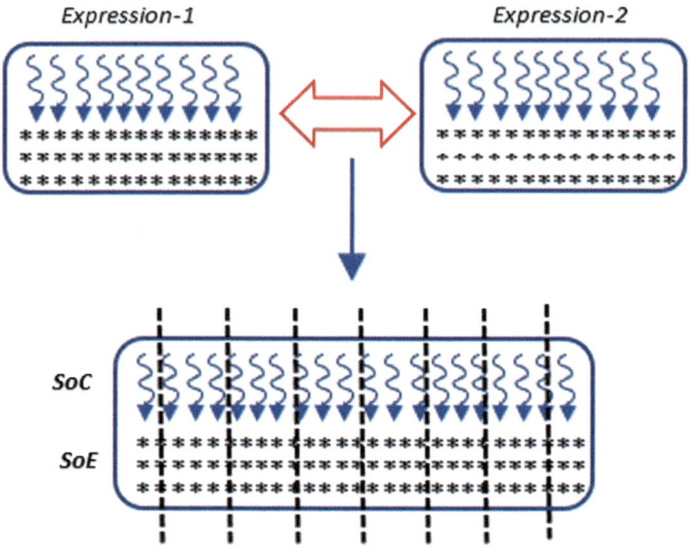

Fig. 6.2 From expression to sign

6.5 A Necessary Overreach

6.5.1 Recalls

There would be a kind of tangible certainty of the sign: the indisputable truth of a kind of flagrant factuality called « sign », which imposes itself on the gaze and, consequently, on the reflection of everyone.

This opinion is expressed, for example, by Saussure, particularly when he attributes to the sign the qualitative term "concrete": « The signs that make up language are not abstractions but real objects [...]; signs and their relations are what linguistics studies; they are the concrete entities of our science. » ([3]: 102). Or Merleau-Ponty, who places the fact of the sign at the forefront of semiolinguistic reality: « [...] the central phenomenon of language is in fact the common act of the signifying and the signified [...] » ([24](1960): 95).

Nevertheless, this certainty attached to the fact of the sign, to which we must acknowledge the force of evidence, is generally reconsidered, as much by the great authors who affirm it as by many other theoretical perspectives that dissolve its unity. We have seen the example of Saussure, who, even though he initially defines language as "[...] a system of signs in which the only essential thing is the union of meaning and acoustic image [...]" ([3]: 15, emphasis added), further denies the primacy of the sign, first by allocating this primacy to the system: « [...] to consider a term as simply the union of a certain sound with a certain concept is grossly misleading. To define it in this way would isolate the term from its system; it would mean assuming that one can start from the terms and construct the system by adding them together when, on the contrary, it is from the interdependent whole that one must start and through analysis obtain its elements. » ([3]: 113). Then, as we have seen (cf. supra), Saussure rejects the sign in its very "essence" (cf. supra). Similarly, Merleau-Ponty reduces the « factual density » of the sign by allocating to it only a transitory existence: a simple point of passage between, on the one hand, fundamental diacritical forms (here Merleau-Ponty takes up and reinvents Saussurean differentiality) and, on the other hand, the outcome of an operation of "consummation" which, leading to a counterpart of the sign, correlatively annihilates its existence.

Indeed, let us recall that, reinvesting Saussurean thought, Merleau-Ponty situates the differential power of phonemes at the foundation of speech and of the act of signifying, and thus below the phenomenon of the sign: Since difference is the mode of existence of signification, and « each [sign] signifies only its difference in respect to the others [...] » ([24]: 88), it is therefore in these linguistic elements devoid of signification and which function differentially that we should locate the original form and essence of the "verbal gesture". These are the phonemes, "oppositional, relative and negative" identities, as « [...] components of language which do not for their part have any assignable meaning, [that have the] function to make possible the discrimination of signs in the strict sense » ([24]: 40). The phonemes, which deliver « an inexhaustible power of differentiating one linguistic gesture from another » ([22]: 33), thus constitute the « [...] real foundations of speech, since they [...] by themselves mean nothing one can specify. But for this very reason, they represent the originary form of signifying. They bring us into the presence of that primary operation, beneath institutionalized language, that creates the simultaneous possibility of significations and discrete signs. » (ibid.: 33). Below constituted language, it is therefore in phonological systems that we can locate « This primordial level of language [...] by defining signs, as Saussure does, not as the representations of certain significations but as the means of differentiation in the verbal chain and [...] in speech. » (ibid.: 31).

But if the sign, in its common sense of undivided unity of a signifier and a signified, loses its primary position in the system of language and in the act of speech, it does not acquire the status of an ultimate term, of an outcome. For, as Merleau-Ponty insists, among the powers of language, and even as a higher form of its accomplishment, is that of leading to universes of ideas or things, provided with an order of their own, and situated beyond languages. For it is a fact that in their living use « [...] the signs are immediately forgotten; all that remains is the meaning. The perfection of language lies in its capacity to pass unnoticed. But therein lies the virtue of language: it is language which propels us toward the things it signifies. In the way it works, language hides itself from us. Its triumph is to efface itself and to take us beyond the words to the author's very thoughts » (ibid.: 10). This operation, in which the sign vanishes in favour of an ideal or designated counterpart, and remains only as an inert envelope, is called « consummation » (or « fullfilment »): « The power of language [gives] us the illusion of going beyond all speech to things themselves » (ibid.: 41), so « [...] there is still, in the exercise of language, the consciousness of saying something, the presumption of a fulfillment of language [...] » (ibid.: 40). It should be noted that one stage in this process of consumption is rendered in the framework of model theory, where the signifier, having an autonomous position (formal symbol) and being conventionally attached to an object (concept or referent), is able to refer to it at its own expense.

Merleau-Ponty and Saussure, who, although they ultimately reduce the sign to the status of a transitory state—a fleeting, evanescent reality in the all-encompassing flow of semiotic activity—nevertheless primitively concede it a certain positive truth, and thus set the sign, if only temporarily, at the center of their theoretical approach. Thus, for Merleau-Ponty, Confer the thesis, discussed at length in PhP, that "the word has a meaning", a thesis that establishes a problematic of expressivity and, at the same time, opens up a questioning of the logic and possibility of a reciprocal incorporation of "form" (understood here as the perceptible face of the sign) and of meaning.

When we examine these two approaches, it becomes clear that the central and "positive" fact of semiolinguistic phenomenality, namely the undivided unity of the signifier and the signified, a fact whose manifest nature could explain the persistence of questions about the sign throughout the ages, is "outdated", or at least dissolved in all-encompassing structures. The same is true of other approaches that take up the sign at the outset, in its blatant form of reciprocal incorporation of a signifier and a signified, and seek to reveal its constitutive principles. This is the case, for example, in Husserl's work, where the sign phenomenon is at the beginning of investigations about intentionality. Now, let us recall that, like Saussure and Merleau-Ponty, and at the end of a long and laborious process of reflection, which progresses through the *Logical Investigations*, the *Ideen* and culminates in the *Leçons sur la théorie de la signification*, Husserl is led to abandon the idea of the sign as a unity per se of a signifier and a signified, in favour of an "external" unity insofar as it proceeds from the all-encompassing structure of the attentional field of consciousness.

6.5.2 Problematical Opening

We can thus observe that in these previous approaches, the sign is systematically reconsidered "downwards", from various angles: its status as a principal phenomenon is relativized; its functional role (for example, as an elementary combinatorial component) is reduced; and its constitutive regimes are no longer attached to it per se, but rather derive from underlying, subsequent or encompassing operations (cf. also the example of glossematics). In short, it is as if the theorizations of semiotic phenomena, turning away from the essence of the sign, only ever deliver adjoining and approximating forms, preparatory or concluding forms.

Considering this singular epistemic conjuncture, the question to be asked is whether this avoidance of the essence of the semiotic fact—which in truth relates to an inability to conceptualize the fact of the sign head-on (cf. above, the question of "internal" and "external" relations), and consequently recognizes only the forms that are adjacent to it, forms that preside over it or through which it is consumed (so, as we have seen, for the logico-algebraic sign as defined in the theory of models) -is to be conceived in terms of approximation, or even limitation: the sign is then like an upper or lower limit of determined operations that administer its formation or dispersion. In this case, the sign could only be approached from its periphery and never grasped at its organizing centre—similar to an irrational value that can always be framed, and as closely as one wishes, by larger and smaller rational numbers. Or, rather than a centripetal vision that approaches the sign from the outside, should we rather defend a centrifugal vision? A vision where the sign, originarily assumed and thought of in its essential character, finds itself, in a kind of obligatory analytical extension, necessarily reconfigured into forms and according to regimes by which it then exceeds what it simply carries within itself, thus acquiring the power to institute universes of meaning that do not yet exist. Here we are in line with Merleau-Ponty's views, according to which authentic living speech accomplishes an innovative significant intention, namely to establish in a verbal body « the excess of what I intend to say over what is being said or has already been said » ([24] : 89).

However, this centrifugal perspective is not without significant difficulties. For, as we have seen (cf. supra), the sign, in its very nature, escapes the categorical forms of a "classical" episteme, such as those used in the empirical sciences. However, the heterogenetic approach, as we have shown, provides an answer to these various problems.

In order to set up a "centrifugal" perspective, which would thus start from the sign thought in the mode of an inner connection of its two sides, to lead to the sign recognized, in conformity with the categorical forms of an empirical objectivity, in its various "approached" operative forms, it is necessary to provide ourselves with a framework of intelligibility of the "interior" relations. To do this, we have the "first" Merleau-Ponty—who, as we know (cf. supra), accounts for the interiorization of bodily (praxical) significations to the sensible qualities of the world of experience through an "elaboration" of the said sensible qualities.

6.6 Inner Relations: Problems and Possible Solution

6.6.1 Problems

However, the Merleau-Pontian approach also has its limitations. For, as it is conceived in the mode of internal relations, the relationship between the organism, as a power of action, and its environment, as a perceived landscape, takes the form of a reciprocal conditioning. And in such a case of strict syntony between the poles of perception and action, the organism finds itself in resonance with its environment, which is none other than the geometry of the actions to be accomplished. Merleau-Ponty uses here (with Scheler) the french word « extase » (ecstasy) : « [the] milieu [...] among which animals live entranced [i.e. in ecstasy]. » (SoB: 176). We are in a situation where there is practically no longer any need to differentiate between body and world, nor to conceive of the relationship of one to the other in terms of action and reaction: « One cannot assign a moment in which the world acts on the organism, since the very effect of this "action" expresses the internal law of the organism. The mutual exteriority of the organism and the milieu is surmounted along with the mutual exteriority of the stimuli. » (SoB: 161).

Undoubtedly, this conjuncture of a strict inter-determination of the world as a geometry of actions and of the body as the driving principle of these actions is too radical, and it is in fact a more or less great distancing of these two planes, of the perceived and the acted upon, which is in fact accomplished (notably in the decoupling of the proper body into its "actual" and the "usual" parts—cf. (PhP : 97 sq.)) Dually, as we have seen (cf. supra), it is in pathological situations that we observe an abolition of the distance between these two planes, which thus induces an obligatory conformation of actions to the environment. Thus, let us recall, discussing these questions, Merleau-Ponty distinguishes between "automatic" or "concrete" behaviours, which therefore adhere to their environment, and "categorical" or "abstract" behaviours, which are the result of a capacity to reconfigure the world and to install new meanings in it (cf. (PhP, 1st Part, Chap. III, p. 114)).

At this point, we find, better established, the questions that were anticipated, namely: How to escape the determinism of the body/world circularity, that is, how to extricate oneself from the gangue of action that constitutes a 'milieu'. In other words, and taking up the terminological options of Merleau-Ponty, how to leave the 'concrete' to reach the creative freedom of the 'abstract'. And this questioning is transposed to the semiotic level in the following way: how to leave the expressive unity which, similarly, determines and encloses the uses of signs, in order to reach the sphere of a free language that creates new meanings? It is easy to imagine that a polarization and a "breaking of symmetry" of the expressive phenomenon will be the key, precisely because it will open up the possibility of overcoming the regimes of legality that administer the order of signs.

6.6.2 Possible Solution

We have thus recalled that "concrete" and "abstract" movements are permeated by two kinds of world-consciousness: concrete movement takes place in a world whose perceived and experienced facture is that of a directive scheme, abstract movement takes place in a world recognized as an "objective milieu": a world of objects free from the power of action of my own body, instituted in themselves, and which in return do not oblige it to anything.

The abstract movement, in its power to produce an environment of new values, would thus coordinate two dimensions: the residual dimension of a world emptied of its former "concrete" meanings, in short the "objective milieu" of a transcendental consciousness, and the dimensions of a projected world of which the "objective milieu", passive and mute, constitutes a possible receptacle.

But it is not only things that are then concerned by this process of objectifying reconfiguration, it is also the gesture, as such: as an engagement of the body with the world, which finds itself installed in an in-between, where, without necessarily losing its spontaneity, it finds itself possibly traced, promoted as a figure, inscribed in an external spatiality, and almost considered objectively in its kinematic physical reality. In this way, the gesture is likely to be reflected upon and thematized in order to be involved in the installation of new milieus.

It is therefore through the effect of a decoupling, which empties the gesture of its originary vital meaning to make it the vector of new meanings, that the subject escapes the conditioning of his environment as perceived and gains access to freedom of action.

The fundamental operation that makes this possible is a kind of dissymmetrization of the perceived in its primary nature as an expressive fact: the phenomenon primitively approached as the presence of a signification is polarized into signifier and signified, and in such a way that the signifier can detach itself from the signified in order to gain its autonomy, but still remain in some way attached to the signified, without which it would completely lose its status as a signifier.

It is this fundamental operation of polarization and dissymmetrization that lies at the foundation of authentic symbolic ability: namely, the capacity to produce meaning, and not to live in the milieu of the mere meanings that originarily the living constitutes for itself as the surrounding world.

For, as Merleau-Ponty insists, « What defines man is not [only] the capacity to create a second nature-economic, social or cultural-beyond biological nature; it is rather the capacity of going beyond created structures in order to create others » (SoB: 175). Now, in order to go beyond the existing structures within which man acts and positions himself as if in a nature, he must possess this power to distance himself from them, thus temporarily neutralizing the meanings established within them, in order to modulate or to abolish them in a second phase in favour of new meanings.

And this is just as true for cultural and symbolic system, whose theorization must then necessarily account for the possibility they offer of contravening their own modes of operation, in other words the possibility of infringing their legislation, in order to institute new structures and new spectra of signification. Thus « [the] use-objects [among which, as Merleau-Ponty points out, are words] and [the] cultural objects would not be what they are if the activity which brings about their appearance did not also have as its meaning to reject them and to surpass them. » (SoB: 176).

Now, as we have seen, this operation presupposes a kind of neutralization of the signified, and a kind of objectification of the signifier; in other words, at the level of the living body and its perceptions, it requires the passage from a "milieu" to a "universe": « [...] perception, which until now has appeared to us to be the assimilation of consciousness into a cradle of institutions and a narrow circle of human "milieus," can become, especially by means of art, perception of a "universe." The knowledge of a truth is substituted for the experience of an immediate reality. » (SoB: 176).

But it is not simply a question of neutralizing the significant dimension of the world as expression, in order to institute in its place an objective universe, no longer belonging to a living consciousness but to a transcendental consciousness. It is also a question of renewing the semantic value of the signifiers that were temporarily neutralized. We can speak here of a reconfiguration of "points of view": "It is this possibility of varied expressions of the same theme, this « [...] "multiplicity of perspective," which is lacking in animal behavior. It is this which introduces a cognitive conduct and a free conduct. In making possible all substitutions of points of view, it liberates the "stimuli" from the here-and-now relations in which my own point of view involves them and from the functional values which the needs of the species, defined once and for all, assign to them. » (SoB: 122).

From this discussion, we will retain that, in order to avoid the problem of "circular causality" between the organism and its environment, and to account for the malleability and evolvability of symbolic systems, we must design semiotic forms that by means of a polarization (into signifier and signified) accompanied by a break in symmetry (between signifier and signified), give a signifier that is possibly detached from its meaning the functional position on which the legality of the semiotic system can be reconsidered in view of establishing new meanings.

To avoid any misunderstanding, it should be made clear from the outset that "polarisation" is not dissociation. Polarization is certainly about making two components appear within an'expressive' unit where « taken in its nascent state [...] the sensible sign and its signification are not even ideally separable. » ([21]: 40-41), but not about « consumming » it, that is to say retaining only an ideal object, after having brought it back to the format of a simple correspondence, conventional at the time, between two autonomous identities, belonging respectively to the sensible and the intelligible. Polarization is not a split but the installation of a tension by which two poles are established, two poles which are then identifiable but which retain a certain form of inter-determination which will have to be made explicit. In short, polarization is an intermediate position between the modes of internal and external links: whereas the terms participating in an internal link contain each other (not really but in terms of meaning), as it is the case for the part with respect to the whole in a holistic perspective, and whereas the terms participating in an external link are each constituted for themselves, and therefore prior to their assembly, the polar connection plays on both pictures, albeit asymmetrically: the poles can be considered, at least by thought, one at a time, and even for one of the two (the signifier) grasped in its material part, but they are not dissociable, insofar as the existence of one is entirely subordinate to that of the other.

It is precisely this polarization of the expressive fact that is realized in the morphodynamics of the Saussurean sign [29–31]. Specifically, in the morphodynamic approach, the substances of expression and content resulting from the heterogeneous process, and which, as constituents of an expressive fact, originarily participate in an inner relation, are now decoupled and then mutually categorized into functional units maintaining relations of asymmetrical conditioning, thus partially reconstituting the lost expressive unity. The signifiers (units of the substance of expression) control the emergence of the signifieds (units of the substance of content) in accordance with the laws of the semiolinguistic system of which they are a part, but they also constitute points of support on which these laws can be overcome. We now present the main elements of the morphodynamics of the sign.

6.7 On the Constitution of the Sign

6.7.1 *Morphodynamics of the Saussurean Sign*

The morphodynamic architecture of the Saussurean sign, in its essentials, is given by the following functional diagram (Fig. 6.3):

This functional architecture articulates two substrate spaces (substance of expression (SoE) and content (SoC)) and a space F of dynamics. Let's look at them in turn.

The substance of the expression is a phonic, graphic or other purport, configured in phonematic, graphematic units (phonemes, graphemes... as "oppositive, negative and relative identities"). These units enter into "syntagmatic" assemblies to constitute "word forms" (or vocables, here noted α, β) which, let us emphasize, are not at this stage signifiers insofar as no signified are intrinsically paired with them.

A relation R_s (called by Saussure relation of "signification") conceived as a simple conventional correspondence (cf. Husserl's "indicative" sign, supra) associates the vocables of the SoE with sets of occurrences in substance of content—occurrences which the said vocables, therefore, point to and scan, thus for example a and b.

In this theoretical device, the SoC is a supposedly homogeneous and continuous substrate (the "amorphous mass of confused ideas") and its structural moment is

Fig. 6.3 Functional architecture of the saussurean sign

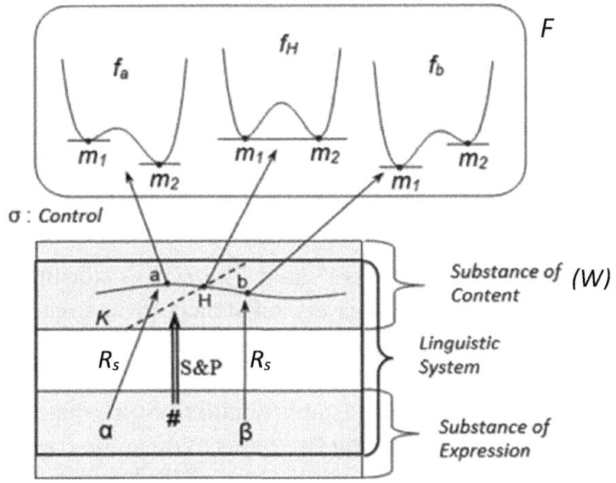

that of its categorization into differential (in the sense of differences) identities (the signifieds) by instantiation of a system of differentiating thresholds (boundaries). As the SoC does not hold the principle of its categorization, the latter requires the involvement of a space of higher complexity whose substance in play will receive structural information in return.

In this respect, the morphodynamic model introduces a space of dynamics F, in functional connection with the substrate space to be categorized (called external space, noted generically W), and within which, therefore, it is a question of explaining the genesis of discontinuities (system of differentiating thresholds) by way of 'inheritance' of the morphological complexity of F. We have already faced the description of the morphodynamic approach in Chap. 2, and we adapt it here to face the semiotic question. In practice, therefore, we "immerse" W in a space of dynamics defined as potential functions with real values, by the action of a field $\sigma : W \to F$ which associates to the units w of W (called control parameters) dynamics f_w of F. In this sense W controls the dynamics of F.

The next step is to study F from the point of view of the spatial distribution of the qualitative types of its elements (potential functions). We must therefore, first define the qualitative type of a potential function, and then, provided with a topology, examine their distribution in this light.

The qualitative types of functions f_w are characterized by the nature and relative positions of their critical points (points in which all first derivatives of f_w are annulled; a critical point is a point in which the tangent to the curve of $f(x)$ is horizontal). In these lines, we will limit ourselves to the case of critical points that are local minima and thus determine stable states of the dynamic system. By convention, the "actual" state of the system f_w is defined by the absolute minimum of the graph associated with it (see Fig. 6.2); the states corresponding to the other (relative) minima are thus understood to be "virtual" (not in the sense of a differential). In the figure above, the elements a, b and H of W determine the dynamics f_a, f_b, f_H, each of which has two minima m_1 and m_2 (attractors in competition for actualization) positioned in different ways (absolute, relative or equal minima).

The f_w dynamics are said to be "stable" if an infinitesimal modification of their "profile" does not modify their qualitative type, that is, if it preserves the relative distributions of the minima. This is the case for f_a and f_b whose attractors m_1 and m_2 preserve their respective positions of absolute or relative minima. Conversely, the potential f_H is unstable because an infinitesimal alteration transforms it into dynamics of the type of f_a or f_b.

The dynamics f_H associated with H thus constitutes a singularity that categorizes the continuum W that is the substance of the content: along a path from point a to point b, the competitive configurations (between attractors) that characterize the dynamics remain unchanged from a to H (excluded), then in H the associated dynamics presents an unstable qualitative type that, in turn, institutes H in the quality of a frontier K separating two types of dynamic configurations, and past this frontier, from H (excluded) to b, the values of W determine qualitatively identical dynamics again. Thus the 'crossing' of K (set of singular points) along a (generic) path of W results in a 'catastrophic' modification (i.e. 'switching' from one type of form to another type of form) of the associated dynamics [37].

Applied to the Saussurean theory of the sign, this model accounts for the constitution of differential identities of meanings within a supposedly homogeneous content substance. Indeed, the control function σ links elements w of the content substance (set of actual meanings) to dynamic configurations considered from the point of view of their qualitative types, and the SoC receives in return the trace, as a network of boundaries, of the instabilities of F separating various qualitative types. In this way, the identities of the SoC are negative and limiting: they proceed from a delimitation of the SoC into mutually adjoining and limiting domains. If we now consider the stratum of vocables, i.e. the SoE, the control σ is preceded by the relation of signification R_s, which then functions as primary control, or pre-control. In this way, through the composition of R_s and σ, the vocables of the SoE are involved in the elaboration of the signifieds as differential identities, and thus acquire the status of signifier.

More precisely, it will be observed that the functional scheme established does justice to the undivided (albeit asymmetric) unity of the sign.

Indeed, the primary and then secondary control relationship, which link signifiers to the dynamics (f_w) that determine differential structurations of the substance of the content, is a relationship that, from one point of view, calls upon units (control parameters) belonging to a (control) space with its own constitutive regimes (e.g., a metric, or, for our purposes, a phonological order), and, from another point of view, functionally establishes these units as "pure" constitutive factors of differential identities belonging to a distinct sphere.

If we focus on this second aspect, i.e., on the morphogenetic function of the signifier, as assigned to it by the morphodynamic device, it is the integrated and dissymmetrical character of the sign as an association of a signifier and a signified that comes to the fore. Indeed, first of all approximately: since the existence of signifieds, as terms of a process of emergence of differential values, is conditioned by signifiers, the latter are therefore by construction "consubstantial" with signifieds. This is an almost analytical truth. But if the signifieds cannot be conceived separately from the signifiers, the same cannot be said of the signifiers themselves: signifiers (as control factors) are, always by construction, a functional prerequisite for the establishment of signifieds, but not vice versa.

Signifiers and signifieds thus appear as the functionally paired, albeit unequal (dissymmetry), poles of the sign. Morphodynamics thus establishes the sign in its integrated unity in a polarized form, which, as discussed above, opens up a way out of "expressive confinement ».

6.7.2 The Autotransgression of the Semiotic

Considering the initial morphodynamic scheme, we must now account for the double arrow "⇒". In doing so, and essentially, it will be a matter of moving from a (simplified) morphodynamics of the isolated sign to a morphodynamics of the interactions between signs, as they are thus established according to the modes of the syntagmatic and the paradigmatic. Let us first observe that the relations of negative difference in content plane—that is, at the level of the signified—condition the very existence of signs: the disappearance of a boundary in the substance of content has the consequence of bringing into continuity, i.e. homogenizing, the two sub-domains (the signifieds) that it institutes according to relations of reciprocal limitation. Such a structural "collapse" thus affects the existence of the signifieds and, at the same time, that of the signs that imply them.

That is, the "⇒" arrow (which governs the installation of boundaries in the substance of content) is functionally involved in the question of existence versus non-existence in language.

Furthermore, it is in the syntagmatic and the paradigmatic as variational axes that the modalities of the existent and the non-existent in language are brought into play and meet. More precisely, the S&P relations, insofar as they administer the variations of a given syntagm, constitute an operative structure that deals with the possible and the impossible in language. This is the case, for example, with differential pairs, which are constantly used in linguistic analysis, and which precisely and methodologically stage the exit from linguistic legality, in other words the exit from the sphere of existence in language. It should be emphasized that we are not dealing here with a global, one-piece legality, but with a local, stratified legality, which makes it possible to conceive of punctual distortions in the form of alterations of thresholds, in a logic of targeted adjustment, reconfiguration and negotiation of meaning in speech.

We shall therefore retain that the relation of determination "⇒", which governs existence and non-existence in language, ultimately refers structurally to the order of S&P relations.

We can thus understand the functional meaning of the sign's dissymmetry. For if the signifier and the signified shared the same status and function, in other words, if they had equivalent roles as constituents of the sign, the annihilation of one would entail the annihilation of the other, and vice versa, and it would then be impossible to imply syntagmatic configurations in language that go beyond linguistic legality, for the purpose of semantic construction. But this is not the case, as is shown by the "maintenance" of the signifier even when no signified is actualized: when the process of content differentiation fails as an echo of a transgression of linguistic legality, thus annihilating all semantic existence in language, the face of the signifier nonetheless remains in linguistic consciousness as phonetic or graphemic complexes, thus opening up to a void of meaning.

The dissymmetry of the sign is thus, in part, the functional correlate of a system that, via S&P relations, and particularly insofar as these latter bring functionally into play that which is impossible in language, incorporates the modalities of its own transgression.

6.7.3 Conclusion

Morphodynamics thus establishes the sign in a polarized form: the faces of the signifier and the signified, even if interdependent (albeit asymmetrically), can be approached and qualified distinctly from one another, precisely in terms (substance and form) that fall within the categoricality of the empirical sciences. In this way, too, morphodynamics places the sign in a structural configuration intermediate between the order of internal and external relations—the former administering the expressive fact, the latter presiding over an empirical objectivity—and in so doing confers on it the power to generate new dimensions of meaning, eliminating thereby the expressive "envelope". For, as we have seen, if the sign polarized in this way makes it possible to "negate and surpass" the grid of available meanings and values, it is because of an inequality of status between signifier and signified. As signifieds are differential identities, and as the signifier controls the emergence of differential relations at the level of the content substances, then administering (via S&P relations) the identity and existence of signifieds, the signifier thus constitutes a fulcrum for the reconfiguration of established meaning(s).

References

1. Benveniste, E.: Problems in general linguistics, M. E. Meek (trans.). University of Miami Press, Coral Gables, FL, 1971 (1966)
2. Cadiot, P., Visetti, Y.-M.: Pour une théorie des formes sémantiques?: motifs, profils, thèmes, PUF, coll. Formes Sémiotiques, Paris (2001)
3. de Saussure, F.: Course in General Linguistics (W. Baskin, Trans.) Philosophical Library, vol. 36. New York, 1959 (1916)
4. de Saussure, F.: Writings in general linguistics. In: Bouquet, S., Engler, R., Sanders, C. (eds.) and Matthew Pires. Oxford University Press, Oxford, 2006 (2002)
5. Deledalle, G: Théorie et pratique du signe: Introduction à la sémiotique de Ch. S. Peirce, Paris: Payot (1979)
6. Eco, U.: Semiotics and Philosophy of Language. MacMillan Press, London (1984)
7. Eco, U.: Sémiotique et philosophie du langage. PUF, Paris (1988)
8. Everaert, N.: Le processus interprétatif: Introduction à la sémiotique de Ch. S. Peirce. Mardaga, Paris (1990)
9. Fontanille, J: Pratiques sémiotiques: Immanence et pertinence, efficience et optimisation, Nouveaux Actes Sémiotiques, pp. 104–106 (2006)
10. Fontanille, J.: Modes du sensible et syntaxe figurative. Actes Sémiotiques **61–63**, 1–69 (1999)
11. Fontanille, J.: Soma et séma: Figures du corps. Maisonneuve et Larose, Paris (2004)
12. Godel, R.: Les sources manuscrites du Course de Linguistique Generale de F. de Saussure, (Publications Romanes et Francaises 61). Droz, Geneve (1969)
13. Hjelmslev, L:,1971. Essais linguistiques, Éditions de Minuit, coll. Arguments, vol. 47. Paris (1908)
14. Hjelmslev, L.: Nouveaux essais, PUF, coll. Formes sémiotiques, Paris (1985)
15. Husserl, E.: Logical Investigations: Prolegomena, Investigations I & II. Translated by J. N. Findlay. London & New-York, Routledge , 2001 (1901)
16. Husserl, E.: Vrin. Lecons sur la theorie de la signification, Paris (1995)
17. Jørgensen, H., Stjernfelt, F.: Substance, substrat, structure. Langages **86**, 79–94 (1987)

18. Kant, I: Critique of Pure Reason (P. Guyer, Trans). Cambridge University Press, 1998 (1787)
19. Lo Piparo, F.: Saussure et les Grecs. Cahiers Ferdinand de Saussure **60**, 139–162 (2007)
20. Merleau-Ponty, M.: Le Monde sensible et le monde de l'expression. Notes du cours au Collège de France, texte établi et annoté par E. de Saint Aubert & S. Kristensen, Genève, MétisPresses (2011)
21. Merleau-Ponty, M.: Phenomenology of perception, D. A. Landes (trans.). Routledge, London, 2012 (1945)
22. Merleau-Ponty, M.: The Prose of World (John O'Neill Trans). Northwestern University Press, Evanston, 1973 (1969)
23. Merleau-Ponty, M.: The Structure of Behaviour (A. L. Fisher, Trans.). Beacon Press, Boston
24. Merleau-Ponty, M.:, 1964 (1960). Signs (R. C. McCleary Trans), Northwestern University Press, 1963 (1942)
25. Milner, J.-C.: Introduction á une science du langage, Le Seuil, coll. Des Travaux, Paris (1989)
26. Peirce, C.S.: Semiotics and significs: the correspondence between Charles S. Peirce and Victoria Lady Welby. C. S. Hardwick, ed. Indiana University, Bloomington, 1977 (1908)
27. Peirce, C.S.: The Collected Papers, Electronic Edition (available on line) (1931–1966)
28. Petitot, J.: Physique du sens? de la théorie des singularités aux structures sámio-narratives. Éditions du CNRS, Paris (1992)
29. Piotrowski, D.: Dynamiques et structures en langue. CNRS Éditions, coll. Sciences du Langage, Paris (1997)
30. Piotrowski, D.: Phénoménalité et Objectivité Linguistiques. Champion, Collection Bibliothéque de Grammaire et de Linguistique, Paris (2009)
31. Piotrowski, D.: Morphogenesis of the Sign. Springer, Berlin (2017)
32. Rosenthal, V., Visetti, Y.-M.: Modéles et pensées de l'expression: perspectives microgénétiques. Intellectica **50**, 177–252 (2008)
33. Rothschild, L., Stein, E.M.: Hypoelliptic differential operators and nilpotent groups. Acta Math. **137**, 247–320 (1976)
34. Tamba-Mecz, I.: La sémantique, PUF, coll. Que sais-je? 655, Paris (1991)
35. Taylor, C.: Act as expression. In: Anscombe, G.E.M., Diamond, C., Teichman, J. (eds.) Intention and Intentionality: Essays in Honour of GEM Anscombe. Cornell University Press, Ithaca (1979)
36. Taylor, C.: Philosophical Papers: Human Agency and Language, vol. 1. Cambridge University Press, Cambridge (1985)
37. Thom, R.: University of Warwick, Structural Stability and Morphogenesis, 1972 (1989)

Chapter 7
Chiusa: Morphodynamic Poetry

7.1 Individuation Between Potency and Form

In these pages, we have immersed ourselves in a forest of dynamics that we have defined as imaginative because they emerge from a continuous re-composition of the virtual. For such dynamics, the conditions of possibility themselves are subject to continual transformation, going beyond the homogeneous becomings in which the virtual remains fixed: those of physics, for example, that are defined by universal equations or those of dynamic structuralism that change only by parametric variation. In opposition to these classical conceptions the question of becoming, the becoming of forms as we've sought to present it, is not governed by laws but by a continuous composition of forces. There is no rule and no mathematics underlying this composition; only the concreteness of the immanent gesture. Forces, however, have to be integrated in order to take form; and so there is, accordingly, a mathematics of the integration of forces. Heterogenesis thus holds together the immanent and concrete gesture of composition with the scientific knowledge of the (dis)integration of heterogeneous forces—Spinoza with Leibniz.

Dynamics of this kind are singular and imply different degrees of actualization, which modulate the relationship between virtual potency and actualized form.

There is, in fact, a differential heterogeneity that can be fully integrated, giving rise to the sublime moment of mystical illumination or to the explosion of an idea. The instant of full integration is the Kairos where all differences harmonise in an organic form. Within this heterogeneity, the space of possibility is global and completely connected, allowing the resolution of any problematic dimension with a fullness of the actual.

But there is also a heterogeneity of forces whose spaces of possibility are only local and in partial conjunctions, like monads vibrating only partially in tune. These are the dynamics of neural assemblies with local geometries that share a common space with a few heterogeneous elements, as in the local construction of the lifted space of Rothschild and Stein. If in Riemannian geometries everything is connected with everything else, sub-Riemannian geometries bring a virtual that can be cut,

© Springer Nature Switzerland AG 2022

A. Sarti et al., *Differential Heterogenesis*, Lecture Notes in Morphogenesis,
https://doi.org/10.1007/978-3-030-97797-9_7

fragmented and that can resist the great binge of the actual. This is the case of Simondonian individuation, a process of only ever partial integration, in which individuation can resist taking shape, often preferring the intensity of potency to the extension of form. At stake is a subtraction from actual becoming, along the lines of the Bartleby's "I would prefer not to", which allows for enduring in potency rather than the compulsion of full actualization.

A more radical kind of heterogeneity allows for the construction of virtual assemblages that open a space of possibility characterized by cuts, islands, and autonomous areas. It is the case of processes of disentanglement from dominant dynamics.

7.2 Disentaglement: Composition Versus Maximization

Imaginative dynamics are not determined by any normative foundation. Their virtual is continuously composed without any aim of maximization whatsoever. If *arkhè* is the principle, the immutable foundation that in physics is the universal differential law, *an-arkhè* is the transformation of the principle by means of an immanent composition of the virtual. Imaginative dynamics have to do with the immanent production of meaning. Also, controlled dynamics can be heterogeneous both in space and in time; they appear, therefore, to be free of constraint, but in fact always optimize a functional. For example, in the case of the economy of capital, what is maximised is profit. In this type of dynamics, the composition plane is colonized in such a way that composition is replaced by the optimization of some variable that is external to the immanence of sense making. It deals with a dominant modality of production that is focused on maximising certain parameters in a completely a-contextual way, exploiting all available resources. Far from being a dynamics without law, such a process is in fact driven by very strict constraints. They can be either abstract dynamics, like those of the economy of finance, or embodied dynamics, in the sense that they become an integral part of the process of subjectivation. For example, the maximization of self-investment, performativity, and permanent self-promotion are part of these optimisation processes. In this case, to return to composition necessitates freeing the individuation process from cognitive automatisms and disentangling it from the constraints of optimization, which is the true great invariant of dynamics of control: "Disentangling our action from the tangles that precede the existence of the will itself—this is the core of what I prefer to call "autonomy" rather than "freedom." [2].

The articulation of imaginative and exploitation dynamics does not concern mental ecology alone but, more generally, all three moments of Felix Guattari's critical ecology: environmental and social, as well as mental. Economic automatisms lead to the full integration of production-consumption systems while leading to the disruption of ecological assemblages as well as of networks of solidarity. Biologists increasingly report anthropogenic disruptions of both organisms and ecosystems, suggesting that these processes are a fundamental, qualitative component of the anthropocene crisis, seemingly generating a fragmentation of the web of life. Confronted with a world increasingly integrated by digital technology and financial abstractions, there is a planet hit by the disruption of life processes.

As André Gorz underlines in one of his most intense passages, if an "ethic of liberation" is conceivable in the era of anthropogenic disruption, it must succeed in subverting the primary causes of the problem: "Ecology does not have all its critical and ethical charge if the ravages of the Earth, the destruction of the natural basis of life are not understood as the consequences of a mode of production, if it is not understood that this mode of production requires maximization yields and uses techniques that violate biological balances. I therefore believe that the critique of the techniques in which domination over men and nature is embodied is one of the essential dimensions of an ethics of liberation." [3]

7.3 Mutant Sensibilities

It is a question of freeing the process of individuation from the dynamics of maximization and exploitation and experimenting with the formation of sensitive and expressive spaces, even beyond the limits of established languages. The capacity of sensitive bodies to feel mutant forms is the condition of possibility of comprehending human and non-human alterity and making it a resource for sense-making.

As we tried to show in the previous pages, sense making is polymorphic because it is grounded on mutant expressive spaces. In such a space, many different things can happen—If structural capture devices are installed, we find the emergence of signs and structures [6]. But there is also the possibility of the dynamics of poetic-literary creation which cannot remain at the level of the recombination of signs but requires reimmersion at the level of a diagrammatic writing which itself is nothing more than a genesis of forms and expressive spaces, even before any given definition of sign [1]. Or the possibility of dynamics without symbolic substitution, such as the primary semiosis that characterizes the relationship between the mother and the newborn baby [8].

More generally, all these processes are founded on embodied dynamics and on the invention of new spaces of possibility that define the expressive character of perception as a critical dimension of life. The logocentric and anthropocentric stance that recognizes the imaginative character of linguistic articulations alone must be overcome in favour of an understanding of an extended imagination that emerges from any form of embodied recomposition of the virtual. Perception itself is already a form of creation. It is at the point of intersection of expressive spaces that the web of life is formed, as the diagrammatic writing of an extended plane of nature. It is in this fusion point that a form of ethics can be defined as proximity to the other in forms of solidarity in which the perception of a community, a territory, a shared destiny, is aimed at the search for a common future. It is useful to reiterate that these expressive spaces go far beyond the automatism of empathy since heterogenesis deals with singular imaginative and historical processes and not with physiological automatisms. Recognizing the imagination of the living means respecting its singularity and asserting the exit from automatism. This process implies a "becoming-with" involving the entire web of life [4].

It's for this reason that the analogy proposed by the British anthropologist Tim Ingold, between polyphonic music and the flow of life, in which "the life of every creature is equivalent to a melody in counterpoint" is particularly apt [5]. Following Ingold, organisms must be understood as bundles of lines—never intersecting, but playing one with another with the sole aim of prolongating themselves:

"Thus in life as in music or painting, in the movement of becoming—the growth of the organism, the unfolding of the melody, the motion of the brush and its trace – points are not joined so much as swept aside and rendered indiscernible by the current as it flows through. So it is that the line does not link the spider and the fly, or the wasp and the orchid, but "passes between them, carrying them away in a shared proximity in which the discernibility of points disappears" (Deleuze and Guattari 2004, p. 324). Life is open-ended: its impulse is not to reach a terminus but to keep on going. The spider spinning its web or the musician launching into the melody "hazards an improvisation." But to improvise, Deleuze continues, is "to join with the World, or meld with it." [5]

Even if we don't, at this point, know how, we know we must rethink the entire ecology of relations in which we live and transmute it in an ecology of transformation that would free our ways of thinking, feeling, and desiring. This is precisely the aim of the nomadic science of material forces: to envisage new compositions of sensitive assemblages as well as the disentanglement of subjects from dynamic automatisms.

What is evoked here is the necessity to see and to feel transformations, making the body a sort of laboratory in which attention must be paid to the continuous variations of differential intensities. The minor science of forces and differentials is qualified as inseparable from a sensible intuition of variation. The transition from quantitative big science to minor, morphological, intensive, qualitative science has as a necessary condition the enhancement of the subject's gaze as well as his knowledge of the dynamics from which he himself emerges in the interaction with human and non-humans agents. It is not a question of rejecting the divergent dynamics of classical structuralism, but of perceiving the constitution of bifurcations and the possibility spaces proper to them even before divergent dynamics are instantiated. To know does not mean only to distinguish, to know means to enjoy the ineffable.

It is a question of using science in an anti-algorithmic sense in order to open perception to sensible variation, thereby modifying it, instructing the gaze and expanding the subject's possibilities for transformation. A new alliance between sciences and aesthetics calls for the experimentation of new forms of extended individuation.

New forms of individuation endowed with a temporality and a feeling of transformation that escape the fetishism of commodities and privilege sensitive experimentation over the automatism of production, political ecology over financial economy, immanent becoming over the dynamics of maximal exploitation.

7.4 The Letter of the Seer

"... I say one must be a seer, make oneself a seer.

The Poet makes himself a seer by a long, gigantic and rational derangement of all the senses. All forms of love, suffering, and madness. He searches himself. He exhausts all poisons in himself and keeps only their quintessences. Unspeakable torture where he needs all his faith, all his super–human strength, where he becomes among all men the great patient, the great criminal, the one accursed—and the supreme Scholar!–Because he reaches the unknown! Since he cultivated his soul, rich already, more than any man! He reaches the unknown, and when, bewildered, he ends by losing the intelligence of his visions, he has seen them. Let him die as he leaps through unheard of and unnamable things: other horrible workers will come; they will begin from the horizons where the. other collapsed! [...] Therefore the poet is truly the thief of fire.

He is responsible for humanity, even for the animals; he will have to have his own inventions smelt, felt, and heard; if what he brings back from down there has form; if it is formless, he gives formlessness. A language must be found. Moreover, every word being an idea, the time of a universal language will come! One has to be an academician—deader than a fossil—to complete a dictionary in any language whatsoever. Weak people would begin to think about the first letter of the alphabet, and they would soon rush into madness!

This language will be of the soul for the soul, containing everything, smells, sounds, colours, thought holding on to thought and pulling. The poet would define the amount of the unknown awakening in his time the universal soul: he would give more—than the formulation of his thought, than the annotation of his March towards Progress! Enormity becoming normal, absorbed by all, he would really be a multiplier of progress!

This future will be materialistic, as you see—Always filled with Number and Harmony, these poems will be made to endure. Fundamentally, it would be Greek poetry again in a new way.

Eternal art would have its functions, since poets are citizens. Poetry will not lend its rhythm to action, it will be in advance. These poets will exist. When the endless servitude of woman is broken, when she lives for and by herself, man—heretofore abominable—having given her release, she too will be a poet! Woman will find some of the unknown! Will her world of ideas differ from ours?—She will find strange, unfathomable, repulsive, delicious things; we will take them, we will understand them.

Meanwhile, let us ask the poet for the new—ideas and forms. All the clever ones will soon believe they have satisfied the demand—it is not so!" [7].

References

1. Batt, N.: Dynamique diagrammatique et hétérogenèse, Séminaire Dynamiques Post-structurelles EHESS (2021)
2. Berardi, F: Freedom and potency. E-flux. https://www.e-flux.com/journal/116/378694/freedom-and-potency/ (2021)
3. Gorz, A.: Ecologica. Galilée, Paris (2007)
4. Haraway, D.: Staying with the Trouble, Duke University Press (2016)
5. Ingold, T.: Point, Line and Counterpoint: From Environment to Fluid Space, in Neurobiology of Umwelt. A. Berthoz, Yves Christen editors, Springer (2009)
6. Piotrowski, D.: Morphogenesis of the sign. Springer, Berlin (2017)
7. Rimbaud, A: Letter to Georges Izambard. https://www.mag4.net/Rimbaud/en/DocumentsE1.html (1871)
8. Violi, P.: How our bodies become us: Embodiment, semiosis and intersubjectivity. J. Cogn. Semiotics IV(1): 57–75 (2009)

Chapter 8
Plates

Heterogeneity	Virtual	Actual
Spatial heterogeneity: geometry	Rothschild and Stein heterogeneous fields X_i	New possibility spaces, Morphogenesis of space
Time heterogeneity	Heterogenous differential operators $A(t)$	Phenotype diversity, Ovid metamorphosis, mutation
Time heterogeneity	Operator morphing $A(t)$	Embodied plasticity, 4E cognition
Space-time heterogeneity: Geometry-dynamics	Heterogenous differential operators $A(x, t)$	Composition-actualisation of percepts
Substance heterogeneity	Harmonic analysis	Percept semiosis Expression/content polarisation

© Springer Nature Switzerland AG 2022

A. Sarti et al., *Differential Heterogenesis*, Lecture Notes in Morphogenesis,
https://doi.org/10.1007/978-3-030-97797-9_8

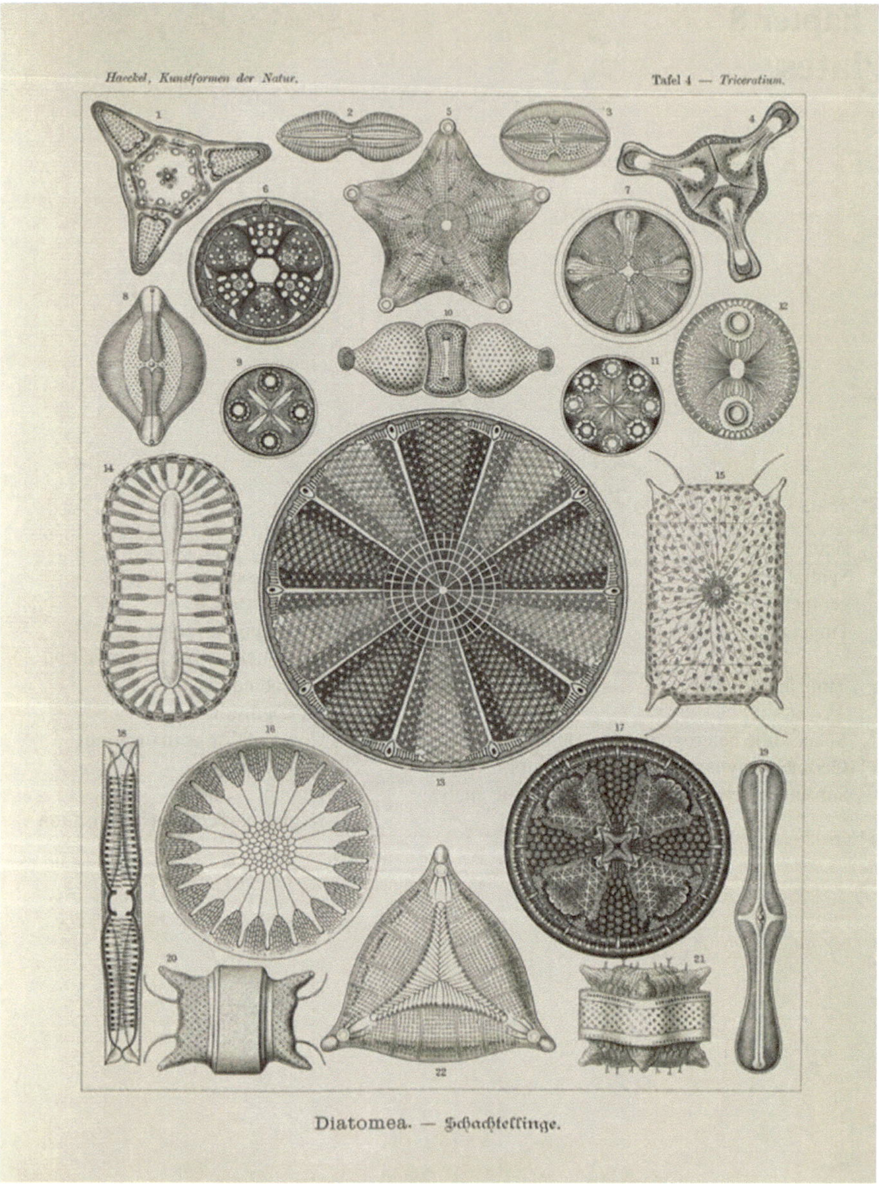

Heterogeneity in phylogenesis: Phenotype diversity.
Ernst Haeckel, Diatomea (Wikipedia)

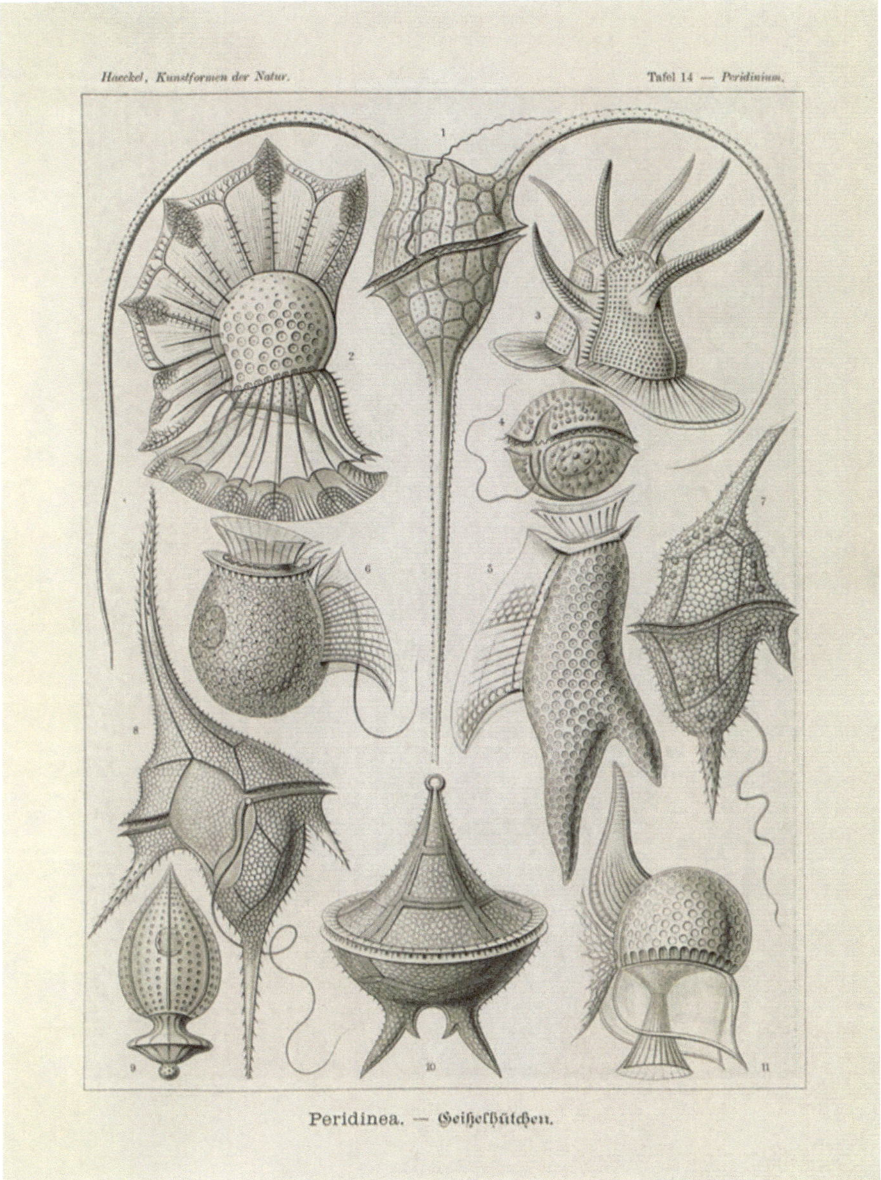

Heterogeneity in phylogenesis: Phenotype diversity.
Ernst Haeckel, Perdinea (Wikipedia)

Heterogeneity in phylogenesis: Phenotype diversity.
Ernst Haeckel, Cyrtoidea (Wikipedia)

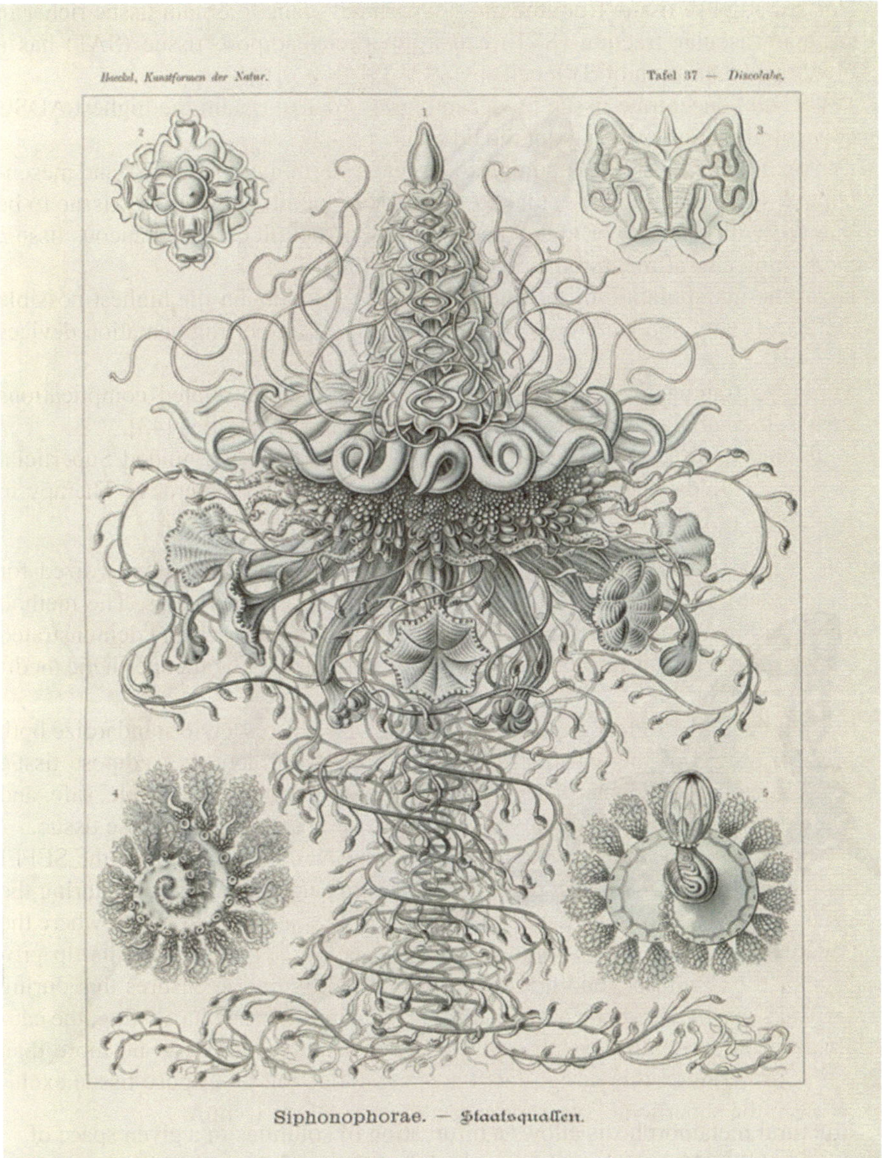

Heterogeneity in phylogenesis: Phenotype diversity.
Ernst Haeckel, Syphonophorae (Wikipedia)

XCIII

Structural metamorphosis allows a bifurcation of solutions in a given space of possibility, but not the evolution of the space itself. The caterpillar can become a butterfly but never become a wolf, because becoming a wolf is not in his space of possibility.
Benjamin Wilkes, Butterfly Metamorphosis

Ovid metamorphosis allows the creation of new spaces of possibility.
Piero del Pollaiolo, Apollo and Daphne (Wikipedia)

Time heterogeneity: Mutations.
Blu, Muto

Time heterogeneity: Mutations.
Blu, Muto.

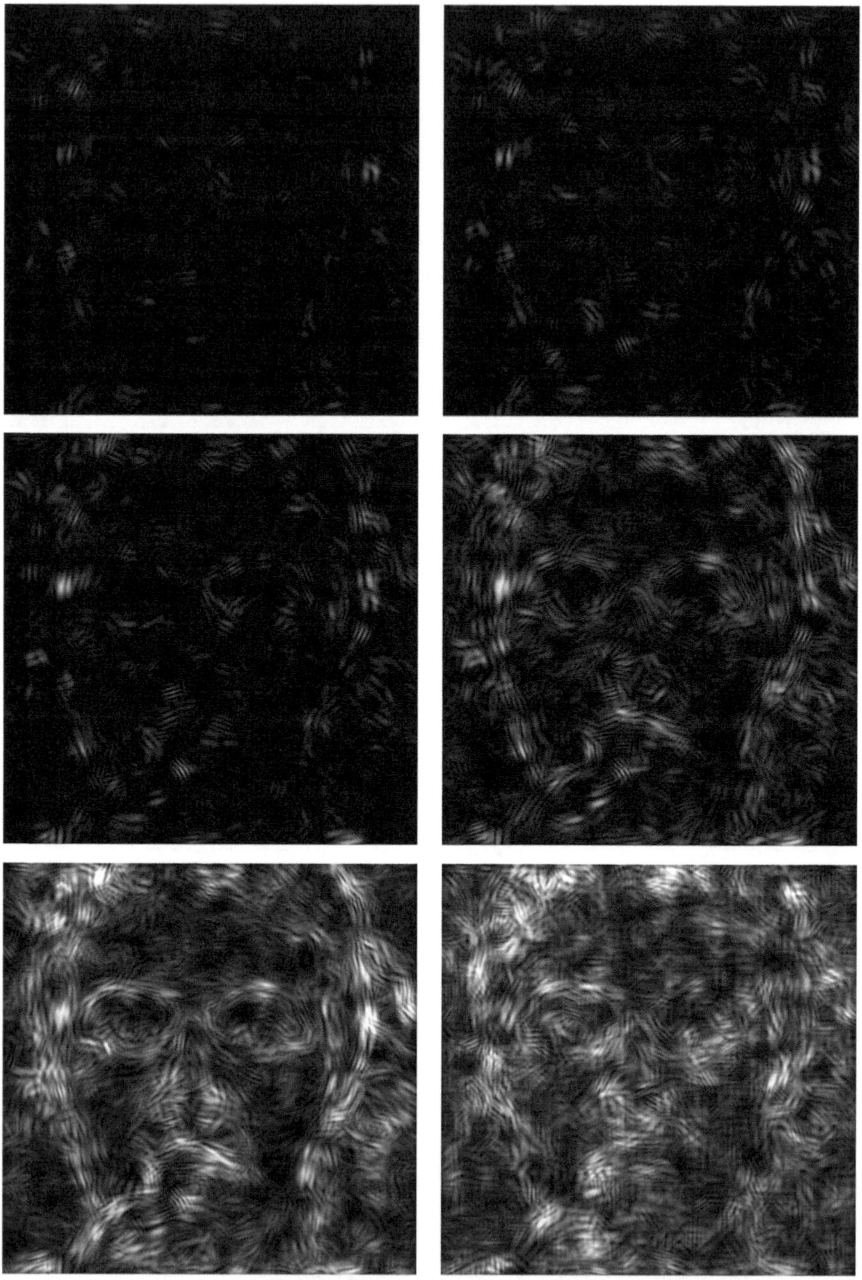

Space-time heterogeneity: percept composition-actualizatione
on the two tim axis of Aion and Khronos.
Jean-Luc Godard, Breathless, 1960.

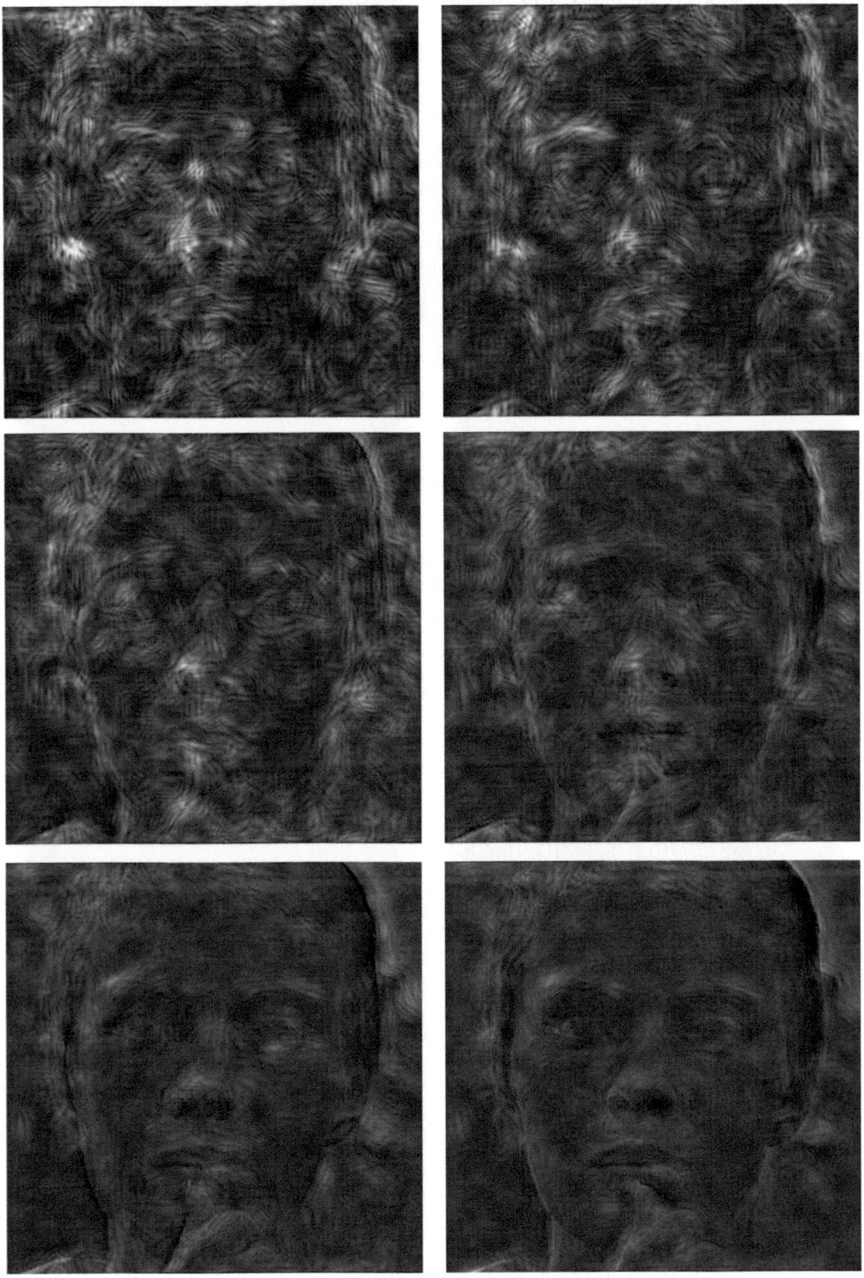

Space-time heterogeneity: percept composition-actualization
on the two time axis of Aion and Khronos.
Jean-Luc Godard, Breathless, 1960.

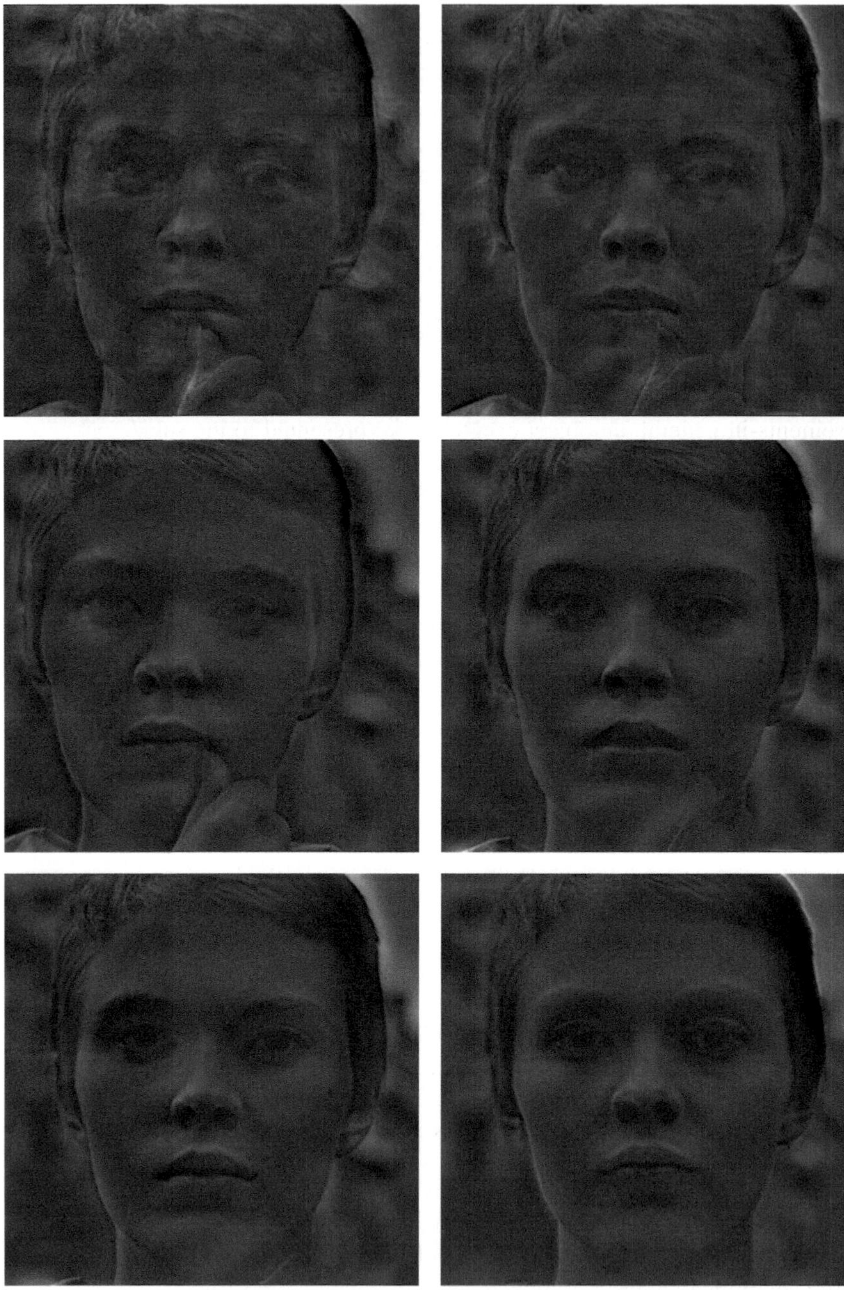

Space-time heterogeneity: percept composition-actualization
on the two time axis of Aion and Khronos.
Jean-Luc Godard, Breathless, 1960.

Subject Index

© Springer Nature Switzerland AG 2022

A. Sarti et al., *Differential Heterogenesis*, Lecture Notes in Morphogenesis,
https://doi.org/10.1007/978-3-030-97797-9

Author Index

© Springer Nature Switzerland AG 2022

A. Sarti et al., *Differential Heterogenesis*, Lecture Notes in Morphogenesis,
https://doi.org/10.1007/978-3-030-97797-9

GPSR Compliance

The European Union's (EU) General Product Safety Regulation (GPSR) is a set of rules that requires consumer products to be safe and our obligations to ensure this.

If you have any concerns about our products, you can contact us on ProductSafety@springernature.com

In case Publisher is established outside the EU, the EU authorized representative is:

Springer Nature Customer Service Center GmbH
Europaplatz 3
69115 Heidelberg, Germany

Batch number: 10726060

Printed by Printforce, the Netherlands